Algebra and Analysis

Algebra and Analysis

Proceedings of the International
Centennial Chebotarev Conference held in
Kazan, Russia, June 5–11, 1994

Editors

M. M. Arslanov
A. N. Parshin
I. R. Shafarevich

W DE G

Walter de Gruyter · Berlin · New York 1996

Editors

M. M. Arslanov
Dept. of Mathematics
Kazan State University
420008 Kazan
Russia

A. N. Parshin
Steklov Math.. Institute
ul. Vavilova 42
117966 Moscow
Russia

I. R. Shafarevich
Steklov Math. Institute
ul. Vavilova 42
117966 Moscow
Russia

1991 Mathematics Subject Classification: 03Dxx, 13Axx, 16Dxx, 16Gxx, 20Kxx, 22Exx

Keywords: Lie algebras, Chebotarev density, singular surfaces, Golay codes, polynomial and simple algebras, Turing reducibility and Turing degrees

♾ Printed on acid-free paper which falls within the guidelines of the ANSI to ensure permanence and durability.

Library of Congress Cataloging-in-Publication-Data

International Centennial Chebotarev Conference
(1994 ; Kazan', Russia)
Algebra and analysis : proceedings of the International Centennial Chebotarev Conference held in Kazan, Russia, June 5–11, 1994 / editors, M. M. Arslanov, A. N. Parshin, I. R. Shafarevich.
p. cm.
ISBN 3-11-014803-X (alk. paper)
1. Algebra–Congresses. 2. Mathematical analysis–Congresses. I. Arslanov, M. M. (Marat Mirzaevich) II. Parshin, A. N. III. Shafarevich, I. R. (Igor' Rostislavovich), 1923– . IV. Title.
QA150.I54 1994
512'.1–dc20 95-49558
CIP

Die Deutsche Bibliothek – Cataloging-in-Publication-Data

Algebra and analysis : proceedings of the International Centennial Chebotarev Conference held in Kazan, Russia, June 5–11, 1994 / ed. M. M. Arslanov ... – Berlin ; New York : de Gruyter, 1996
ISBN 3-11-014803-X
NE: Arslanov, Marat M. [Hrsg.]; International Centennial Chebotarev Conference ⟨1994, Kazan'⟩

Printed in Germany. Printing: Ratzlow-Druck, Berlin. Binding: Lüderitz & Bauer, Berlin.
Cover design: Thomas Bonnie, Hamburg.

Preface

This volume contains the proceedings of the International Conference "Algebra and Analysis", which was held in commemoration of the centennial of the eminent Russian mathematician N. G. Chebotarev (1894–1947). The conference took place from June 5 to June 11, 1994 at Kazan State University, Kazan, Tatarstan, Russia.

The meeting was attended by more than 200 registered participants from 10 different countries. It was organized by the Departments of Mathematics of Kazan and Moscow State Universities, the V. A. Steklov Institute of Mathematics (Moscow) and the N. G. Chebotarev Research Institute of Mathematics and Mechanics (NIIMM, Kazan). Financial support was also provided by the Division of Mathematics of the Russian Academy of Science and the High School Ministry of Russia.

The members of the Executive Board of the Organizing Committee were: Ju. G. Konoplev (Chairman, Rector of Kazan University), M. M. Arslanov (Vice-Chairman), F. G. Avchadiev (Vice-Chairman), A. V. Kosterin (Vice-Chairman, Director of N. G. Chebotarev Research Institute), S. M. Skryabin (Scientific Secretary). Members of the Committee were also L. A. Bokut', G. N. Chebotarev, S. B. Cooper, A. M. Elizarov, Ju. B. Ermolaev, Ju. L. Ershov, B. G. Gabdulchaev, Ju. E. Chochlov, A. I. Kostrikin, I. A. Lavrov, A. V. Lapin, D. Ch. Muschtari, S. R. Nasurov, A. N. Parshin, I. I. Sachaev, I. R. Shafarevich, L. N. Shevrin, P. L. Ul'janov, and A. G. Vitushkin.

The preparation of the two main branches of the conference was coordinated by A. I. Kostrikin and I. R. Shafarevich (Algebra), and P. L. Ul'janov and A. G. Vitushkin (Analysis).

This volume reflects part of the actual program of the conference. There were eleven invited speakers, of whom E. P. Dolzhenko, V. Ja. Gutljanskii and A. S. Dgumadil'daev did not contribute a paper to the proceedings. W. L. Baily, Jr., V. D. Chugunov, L. D. Eskin, S. L. Tonkonog, D. A. Jordan, S. M. Skryabin were asked by the editors to submit the paper they read in a contributed papers session of the meeting.

M. M. Arslanov
A. N. Parshin
I. R. Shafarevich

Table of contents

Generalized reductive homogeneous spaces and quasi-graded semisimple Lie algebras*

Dmitri V. Alekseevsky and Andrea F. Spiro

Abstract. A Lie algebra ℓ is called graded (resp. quasi-graded) of order k if it is decomposed into a direct sum $\ell = g^{-1} + g^0 + \cdots + g^{k-1}$ of non-zero subspaces g^i such that $[g^i, g^j] \subset g^{i+j}$ for all i, j (resp. for i, j with $i + j \geq -1$). Here $g^{i+j} = \{0\}$ if $i + j \notin \{-1, \ldots, k-1\}$. The quasi-graded Lie algebras of order 1 are exactly the Lie algebras with reductive decomposition.

We prove that any reductive (in particular, semisimple) quasi-graded Lie algebra is direct sum of a quasi-graded Lie algebra of order ≤ 1 and a graded semisimple Lie algebra of order 2. In particular, any simple quasi-graded Lie algebra of order 2 is graded. As a geometric application, we obtain a description of all ν-reductive (in the sense of I. L. Kantor) homogeneous spaces L/L_0 with reductive Lie group L.

1. Introduction

A decomposition

$$\ell = g^{-1} + g^0 + \cdots + g^{k-1} \tag{1.1}$$

of a Lie algebra ℓ into a direct sum of non-zero vector spaces g^i is called a *gradation* if

$$[g^i, g^j] \subset g^{i+j} \quad \forall i, j, \tag{1.2}$$

where $g^{i+j} = \{0\}$ for $i + j \notin \{-1, \ldots, k-1\}$. If (1.2) holds only for i, j with $i + j \geq -1$, we say that the decomposition (1.1) is a *quasi-gradation* of order k. A Lie algebra with a (quasi-)gradation is called *(quasi-)graded.* Note that a quasi-graded Lie algebra ℓ of order 0 is simply a Lie algebra and that a quasi-graded Lie algebra of order 1 is a Lie algebra with a reductive decomposition. A quasi-gradation of any order may be considered as a generalization of the reductive decomposition.

*Partially supported by the MPI 40% and CNR, contract n. 93.00554.CTO1

All gradations of semisimple Lie algebras were classified by S. Kobayashi and T. Nagano [KN]. In this paper, we describe the quasi-gradations of a reductive Lie algebra ℓ. The main result is the following (Theorem 5.1):

Theorem. *Let ℓ be a reductive Lie algebra over $\mathbb{C}$ with a quasi-gradation of order k. Then $k \leq 2$ and, as quasi-graded Lie algebra, ℓ is the direct sum of a reductive quasi-graded Lie algebra a of order less or equal to 1, i.e. of the form*

$$a = a^{-1} + a^{0}, \quad [a^{0}, a^{-1}] \subset a^{-1},$$

where a^0 may also be trivial, and of a semisimple graded Lie algebra

$$b = b^{-1} + b^{0} + b^{1}$$

of order 2. In particular, any quasi-gradation of order 2 of a simple Lie algebra ℓ is a gradation.

We also describe the structure of reductive quasi-graded Lie algebras of order 1, that is of reductive Lie algebras with reductive decomposition (Theorem 3.1).

Quasi-graded Lie algebras naturally arise in connection with the geometry of homogeneous spaces ([Ta], [Ka]) and G-structures ([Al]).

Let $M = L/L_0$ be a homogeneous space of a connected Lie group L and let (ℓ, ℓ_0) be the associated infinitesimal object, where $\ell = \operatorname{Lie} L$ is the Lie algebra of vector fields on M and $\ell_0 = \operatorname{Lie} L$ is the stability subalgebra. The subalgebra ℓ_0 defines a geometric filtration

$$\ell \supset \ell_0 \supset \ell_1 \supset \cdots \supset \ell_{k-1} \supset \ell_k = \{0\}, \quad [\ell_i, \ell_j] \subset \ell_{i+j}$$

of the Lie algebra ℓ, where each subspace ℓ_i consists of vector fields $X \in \ell$, which vanish at the point $o \in eL_0$ together with all derivatives up to order i. The number k is called *order* of the homogeneous space M. If the group L is reductive, the order of $M = L/L_0$ does not exceed 2.

One can associate with the filtered Lie algebra ℓ a graded Lie algebra

$$\operatorname{gr} \ell = g^{-1} + g^{0} + \cdots + g^{k-1} \stackrel{\text{def}}{=} \ell/\ell_0 + \ell_0/\ell_1 + \cdots + \ell_{k-1}/\{0\}.$$

In general, the filtered Lie algebra ℓ is not isomorphic to the associated Lie algebra $\operatorname{gr} \ell$. If it is isomorphic, the homogeneous space $M = L/L_0$ is called *flat* and, if ℓ is semisimple, it is called *flat semisimple homogeneous space* (in the sense of N. Tanaka).

Note that the flat homogeneous space $M = L/L_0$ can be locally identified with the vector space $V = g^{-1}$. Indeed, we can write

$$\ell = g^{-1} + g^{0} + \cdots + g^{k-1} = V + \ell_0, \quad V \cap \ell_0 = \{0\},$$

where $V = g^{-1}$ is a commutative subalgebra and $\ell_i = g^{i} + \cdots + g^{k-1}$, for all i. Hence, the subgroup of L that is associated with V acts, locally, simply transitive on M .

The flat semisimple homogeneous spaces $M = L/L_0$, associated with semisimple graded Lie algebras, have a canonical G-structure, where $G = \mathrm{Ad}_{\ell/\ell_0}(L_0)$ is the isotropy group, and this G-structure can be characterized as the canonical maximal homogeneous G-structure (cfr. [Al]).

We remark that the problem of constructing the isomorphism $\ell \simeq \mathrm{gr}\,\ell$ reduces to a choice of a commutative subalgebra V of ℓ, which is complementary to ℓ_0 and such that

$$\ell = g^{-1} + g^0 + \cdots + g^{k-1}, \tag{1.3}$$

where $g^{-1} = V$ and the subspaces g^i are defined inductively by

$$g^i = \{X \in \ell_i, \quad [V, X] \subset g^{i-1}\}. \tag{1.4}$$

If we drop the condition that the subspace V is a commutative subalgebra, we obtain a quasi-gradation of the Lie algebra depending on V (cfr. §2.1).

According to I. L. Kantor, the homogeneous space $M = L/L_0$, corresponding to a quasi-gradation (1.3), (such that $\ell = \mathrm{Lie}\, L$, $\ell_0 = g^0 + \cdots + g^{k-1} = \mathrm{Lie}\, L_0$) is called a *$k$-reductive homogeneous space*. Note that a simply connected 0-reductive homogeneous space is identified with a Lie group L and a 1-reductive homogeneous space is a reductive homogeneous space in the classical sense. Kantor stated the problem of classification of k-reductive homogeneous spaces. As an immediate corollary of our main theorem (which may be directly extended to the case of Lie algebras over $\mathbb{R}$, by considering their complexifications), we obtain the following description of k-reductive homogeneous spaces of a reductive Lie group.

Theorem. *Let $M = L/L_0$ be a simply connected k-reductive homogeneous space of a connected reductive Lie group L. Then $L = A \times B$ is a direct product of normal subgroups A, B, and M, as homogeneous space of the Lie group L, is isomorphic to*

$$M = A/A_0 \times B/B_0,$$

where A/A_0 is a reductive homogeneous space of the Lie group A and B/B_0 is a flat homogeneous space of the semisimple Lie group B.

Remarks. 1) $\mathrm{order}(A/A_0) = 1$ if $A_0 \neq \{e\}$ and it is equal to 0 if $A_0 = \{e\}$. The order of B/B_0 is 2.

2) The list of all flat semisimple homogeneous spaces B/B_0 is given in [KN] and [Oc].

Corollary. *Any 2-reductive homogeneous space L/L_0 of a simple Lie group L is flat.*

2. Definition and first properties of quasi-graded Lie algebras

2.1. Definitions

Definition 2.1. Let ℓ be a finite-dimensional Lie algebra and let ℓ_0 be a proper subalgebra of ℓ. Then, ℓ_0 is called *effective subalgebra* of ℓ if there exists no ideal of ℓ which is contained in ℓ_0. In this case, the pair (ℓ, ℓ_0) is said to be an *infinitesimal homogeneous space.*

Remark 2.2. Let (ℓ, ℓ_0) be an infinitesimal homogeneous space. Consider the connected and simply connected Lie group $\mathcal{L}$ having ℓ as Lie algebra and suppose that $\mathcal{L}_0$ is the connected Lie subgroup corresponding to ℓ_0. The pair (ℓ, ℓ_0) will be called *regular* if the subgroup $\mathcal{L}_0$ is closed in $\mathcal{L}$: in such a case, the coset space $\mathcal{L}/\mathcal{L}_0$ has a structure of differentiable manifold, i.e. it is an homogeneous space and it is canonically associated to the pair (ℓ, ℓ_0). This motivates our choice for the name "infinitesimal homogenous space" to the pair (ℓ, ℓ_0).

Even with the "non-regular" infinitesimal homogeneous spaces, it is possible to associate a differentiable manifold, but this manifold *is just locally homogeneous, it is not locally equivalent to any globally homogeneous manifold* and it is not canonically defined (cfr. [LT], [Sp]).

However, neglecting for the moment the pathological cases which correspond to the non-regular pairs, for what concerns further applications, we mainly consider the *regular* infinitesimal homogeneous spaces and their relation with homogeneous manifolds.

Terminology. If ℓ is a topological Lie algebra (even not finite-dimensional), with linearly compact topology, any subalgebra ℓ_0, with the properties of Definition 2.1, is also called *fundamental subalgebra*, and ℓ is called *transitive Lie algebra* ([Gu]). Note that any finite-dimensional Lie algebra ℓ is linearly compact in the discrete topology; so, in the case of finite dimension, effective subalgebras are the same of fundamental subalgebras.

Any infinitesimal homogeneous space (ℓ, ℓ_0) is naturally associated with a *filtration*, i.e. a descending sequence of subalgebras ℓ_i of ℓ such that $[\ell_i, \ell_j] \subset \ell_{i+j}$ for all integers i, j. In fact, we may define

$$\ell_k = \ell \qquad \forall k \leq -1,$$

$$\ell_1 = \{\xi \in \ell_0 \mid [\ell, \xi] \subset \ell_0\},$$

$$\ell_2 = \{\xi \in \ell_1 \mid [\ell, \xi] \subset \ell_1\},$$

$$\vdots$$

$$\ell_i = \{\xi \in \ell_0 \mid [\ell, \xi] \subset \ell_{i-1}\}.$$

Being ℓ_0 effective, this construction implies that $\ell_i \neq \ell_{i+1}$ for all $i \geq -1$, unless $\ell_i = \{0\}$.

Definition 2.3. The sequence ℓ_i is called the *(geometric) filtration associated with the infinitesimal homogeneous space* (ℓ, ℓ_0).

The first integer i such that $\ell_i = \{0\}$ is called *order* of the transitive Lie algebra (ℓ, ℓ_0) and it coincides with the cardinality of the family of non-trivial spaces ℓ_i/ℓ_{i+1}.

Remark 2.4. The concept of *geometric filtration* and of *order* of an infinitesimal homogeneous space (ℓ, ℓ_0) has an important geometric interpretation.

Consider a regular infinitesimal space (ℓ, ℓ_0) and let M be the homogeneous manifold $M = L/L_0$, as in Remark 2.2. The Lie algebra ℓ of L is faithfully represented as the algebra of "infinitesimal transformations" of L, i.e. of those complete vector fields, whose flow generates a 1-parameter subgroup of transformations in L. Now, fix a point $p \in M$, say eL_0. Then, ℓ_0 is faithfully represented as the subalgebra of vector fields in ℓ which vanish at p; any other subalgebra ℓ_i is represented as the subalgebra of vector fields which vanish at p together with the first i derivatives at p.

Within such interpretation, $\mathrm{order}(\ell, \ell_0) = k$ means that a vector field vanishes at p up to order k if and only if it is trivial everywhere. It also means that any two vector fields coincide at p up to order k if and only if they are equal everywhere.

To consider an example, recall that if M is a homogeneous Riemannian manifold, the infinitesimal transformations of the group $\mathrm{Iso}(M)$ of isometries are the Killing vector fields on M. In this case, the infinitesimal homogeneous space is of order at most 1: in fact, by a well known property of Killing vector fields, two such vector fields coincide everywhere if they coincide up to order 1 at a point $p \in M$.

Lemma 2.5. *If (ℓ, ℓ_0) is an infinitesimal homogeneous space and ℓ is reductive, then*

$$\mathrm{order}(\ell, \ell_0) \leq 2.$$

Proof. Suppose not and let $X \neq 0$ be an element in ℓ_i with $i \geq 2$. For any other element Y in ℓ, the operator $\mathrm{ad}_\ell(X)\,\mathrm{ad}_\ell(Y)$ is nilpotent in $gl(\ell)$: in fact it sends any subalgebra ℓ_k into the subalgebra $\ell_{(k+i-1)} \subset \ell_{(k+1)}$. If we denote by $\mathcal{B}$ the Cartan–Killing form of ℓ, then $\mathcal{B}(X, Y) = 0$ for any $Y \in \ell$, and, hence, X is in the center of ℓ. So, $\mathbb{R}X$ is a non-trivial ideal of ℓ and this contradicts the definition of effective subalgebra. □

Definition 2.6. 1. Let ℓ be a Lie algebra. A *quasi-gradation* of ℓ is a family of subspaces g^i, $-1 \leq i \leq k-1$, such that:

a) ℓ equals, as vector space, the direct sum $\ell = g^{-1} + g^0 + g^1 + \cdots + g^{k-1}$;

b) for any i, j such that $i + j \geq -1$,

$$[g^i, g^j] \subset g^{i+j}$$

(we assume the convention $g^i = \{0\}$ for any integer $i < -1$ or $i \geq k$).

The family g^i is called a *gradation* (or *flat quasi-gradation*) if b) holds even when $i = j = -1$, i.e. if $[g^{-1}, g^{-1}] = \{0\}$.

2. A *quasi-graded Lie algebra* is a pair which is formed by a Lie algebra ℓ and a fixed quasi-gradation of it. A quasi-graded Lie algebra will be called *flat* (or simply *graded*) if the quasi-gradation is indeed a gradation.

3. The *order* of a quasi-gradation $\{g^i\}_{-1 \leq i \leq k-1}$ is the integer k.

Remark 2.7. For any infinitesimal homogeneous space (ℓ, ℓ_0) and any complementary subspace V to ℓ_0 in ℓ (i.e. a subspace such that $\ell = V + \ell_0$), we may define the following sequence of linear subspaces:

$$g^{-1} = V, \quad g^i = \{A \in \ell_i \mid [A, V] \subset g^{i-1}\} \quad \forall i \geq 0.$$

Consider now the subspace

$$\ell(V) = V + g^0 + g^1 + \cdots + g^{k-1}$$

($k = \operatorname{order} \ell$).

We have the following fact:

Lemma 2.8. *For any infinitesimal homogeneous space (ℓ, ℓ_0) and any complementary subspace V to ℓ_0, if $[V, V] \subset \ell(V)$, then $\ell(V)$ is a subalgebra and it is quasi-graded, i.e. for any i, j such that $i + j \geq -1$,*

$$[g^i, g^j] \subset g^{i+j}.$$

Proof. It is proved by induction on the integer $n = i + j$ and by the Jacobi identities. □

Observe that, in a generic case, $\ell(V)$ is a proper subspace of ℓ. But we want to single out the case when $\ell(V)$ coincides with ℓ.

Definition 2.9. Let (ℓ, ℓ_0) be an infinitesimal homogeneous space. We say that *a complement V to the effective subalgebra ℓ_0 generates a quasi-gradation of (ℓ, ℓ_0)* if $\ell(V) = V + g^0 + \cdots + g^{k-1} = \ell$. Note that in this case, for each i, $\ell_i = g^i + \cdots + g^{k-1}$.

Vice versa, we say that *a quasi-gradation $\ell = V + g^0 + \cdots + g^{k-1}$ is effective* if the subalgebra $\ell_0 = g^i + \cdots + g^{k-1}$ is effective and (ℓ, ℓ_0) is an infinitesimal homogeneous space.

When $\ell = V + g^0 + \cdots + g^{k-1}$ is effective, V generates the quasi-gradation, as it is stated by the following immediate lemma.

Lemma 2.10. *Let $\ell = V + g^0 + \cdots + g^{k-1}$ be an effective quasi-graded Lie algebra. Then, $\ell(V) = \ell$ and the geometric filtration is given by the subalgebras*

$$\ell_i = g^i + g^{i+1} + \cdots + g^{k-1}.$$

The order of the quasi-gradation coincides with $\text{order}(\ell, \ell_0)$.

At last, let us name the subalgebras which inherite the quasi-gradation properties.

Definition 2.11. Let $\ell = V + g^0 + \cdots + g^{k-1}$ be a quasi-graded Lie algebra and ℓ' be one of its Lie subalgebras. We say that ℓ' is a *quasi-graded Lie subalgebra* of ℓ if

$$\ell' = \ell' \cap V + \ell' \cap g^0 + \ell' \cap g^1 + \cdots + \ell' \cap g^{k-1}.$$

In particular, we have that ℓ is direct sum of two quasi-graded ideals ℓ' and ℓ'' if and only if

$$\ell = \ell' \oplus \ell'', \quad V = V \cap \ell' + V \cap \ell'',$$

$$g^i = g^i \cap \ell' + g^i \cap \ell'' \quad \forall i \geq 0.$$

2.2. Basic properties of quasi-graded Lie algebras

In all the following, we denote by $\mathcal{B}$ the Cartan–Killing form of a Lie algebra ℓ, defined over $\mathbb{R}$ or $\mathbb{C}$.

Proposition 2.12. *For a quasi-graded Lie algebra $\ell = V + g^0 + \cdots + g^{k-1}$, the following properties hold:*

1. $[V, V] \subset C_\ell(g^i)$ $i \geq 1$ *($C_\ell(g^i)$ = centralizer of g^i in ℓ);*
2. $\mathcal{B}(g^i, g^0 + g^1 + \cdots + g^{k-1}) = 0$ *for any $i \geq 1$.*

Proof. 1. Consider $v, v' \in V$ and $X \in g^i$. Then

$$[[X, v], v'] - [[X, v'], v] \in g^{i-2},$$

by the properties of quasi-gradations. On the other hand, $[[v, v'], X] \in [\ell, g^i] \subset (g^{i-1} + g^i + \cdots + g^{k-1})$ and therefore

$$[[X, v], v'] - [[X, v'], v] = -[[v, v'], X] \in g^{i-2} \cap (g^{i-1} + g^i + \cdots + g^{k-1}) = \{0\}.$$

2. For any element $X \in g^i$ and $Y \in g^0 + g^1 + \cdots + g^k$, the linear map $\text{ad}_\ell(Y)\,\text{ad}_\ell(X)$ sends each module of grade m into a module of grade larger or equal to $m + 1$. Hence $\text{ad}_\ell(Y)\,\text{ad}_\ell(X)$ is nilpotent and $\text{Tr}(\text{ad}(Y)\,\text{ad}(X)) = \mathcal{B}(Y, X) = 0$. □

From Lemma 2.5, it follows immediately the next:

Corollary 2.13. *If the quasi-graded Lie algebra $\ell = V + g^0 + \cdots + g^{k-1}$ is reductive and effective, then*

$$\mathrm{order}(\ell, \ell_0) \leq 2.$$

2.3. Flat reductive quasi-graded Lie algebras

The case where an effective quasi-graded Lie algebra $\ell = V + g^0 + g^1$ is semisimple and flat has been fully explored by S. Kobayashi and T. Nagano in [KN], and there they give a complete list of all simple graded Lie algebras of order 1. Being any graded effective semisimple Lie algebra of order 1, a direct sum of graded simple ideals (Proposition 3.2, [Oc], p. 165), any reductive and effective graded Lie algebra is described as follows:

Theorem 2.14. *Let $\ell = V + g^0 + g^1$ be a reductive graded Lie algebra of order* 1 *with $\ell_0 = g^0 + g^1$ effective, Then:*

1. *the center z of ℓ is contained in V;*
2. *$\ell = z \oplus \bigoplus_{i=1}^n \ell_{(i)}$, where each ideal $\ell_{(i)} = \ell_{(i)} \cap V + \ell_{(i)} \cap g^0 + \ell_{(i)} \cap g^1$ is simple and graded.*

Proof. 1. Since ℓ_0 is effective, $z \cap \ell_0 = \{0\}$. On the other hand, we may write any $X \in z$ in the form

$$X = X_V + X_0 + X_1$$

where $X_V \in V$, $X_0 \in g^0$ and $X_1 \in g^1$. By the gradation properties, it follows that

$$[X, g^i] = 0 \Rightarrow [X_1, V] = [X_1, g^0] = [X_1, g^1] = \{0\}$$

and

$$[X, g^i] = 0 \Rightarrow [X_0, V] = [X_0, g^0] = [X_0, g^1] = \{0\}.$$

Hence, $X_0 + X_1 \in z \cap \ell_0 = \{0\}$ and $X = X_V$.

2. Consider the decomposition $\ell = z + \ell'$ with $\ell' = [\ell, \ell] \simeq \ell / z$. So, for any $X \in \ell$, we may write

$$X = X' + Z \qquad \text{where} \qquad X' \in [\ell, \ell], \quad Z \in z.$$

Since $Z \in V$, whenever $X \in g^0 + g^1$ we have that $Z = 0$; and, if $X \in V$, by the grading properties, it follows that $X' \in [V, g^0]$. So it is clear that

$$\ell' = [V, g^0] + g^0 + g^1 = \ell' \cap V + g^0 + g^1$$

is graded and semisimple. The claim now follows from the mentioned classification of semisimple graded Lie algebras. □

3. Reductive quasi-graded Lie algebras of order 1

We describe here the structure of the reductive quasi-graded Lie algebras of order 1.

Theorem 3.1. *Let ℓ be a reductive Lie algebra and g^0 be a subalgebra of ℓ. Then, there exists a quasi-gradation of order* 1

$$\ell = V + g^0, \quad [g^0, V] \subset V \tag{3.1}$$

having g^0 as isotropy subalgebra, if and only if:

a) *g^0 is a reductive Lie algebra, that is $g^0 = s+z$, where $s = [g^0, g^0]$ is semisimple and z is the center of g^0;*

b) $z \cap [z, \ell] = \{0\}$.

Moreover, the quasi-gradation (3.1) *is unique if and only if there exists no g^0-submodule of V that is isomorphic to a g^0-submodule of g^0.*

Proof. Suppose that $\ell = V + g^0$ is a quasi-gradation of ℓ. Consider a Levi–Malcev decomposition $g^0 = s + r$ of g^0, where r is the radical and s is a semisimple subalgebra. Suppose now that $[g^0, r] \neq \{0\}$ and that X is a non-trivial element of $[g^0, r]$.

Then, the operator $\mathrm{ad}_\ell X$ is nilpotent. By the Morozov–Borel theorem (see Theorem 17 [Ja], p. 100), X is contained into an sl_2-subalgebra of ℓ. In other words, there exist other two elements H and Y in ℓ, such that

$$[H, X] = 2X, \quad [H, Y] = -2Y, \quad [X, Y] = H. \tag{3.2}$$

Moreover, by hypothesis, the subalgebra g^0 has a complementary subspace V invariant under the adjoint action. So, we may also assume (cfr. Lemma 8 [Ja], p. 99) that H, X and Y are elements of g^0. Being $X \in r$, H and Y would also be in r and they would generate an sl_2-subalgebra of the radical: absurd. This contradiction shows that $[g^0, r] = \{0\}$ and proves a).

Moreover,

$$[z, \ell] = [z, V + g^0] = [z, V] \subset V$$

and it has obviously a trivial intersection with z. This proves b).

Conversely, assume that a) and b) hold. Then, if we indicate by $c_\ell(s)$ the centralizer of s, we may write the following decomposition of ℓ into $\mathrm{ad}_\ell(s)$-invariant subspaces:

$$\ell = s + c_\ell(s) + [s, \ell^\perp].$$

(Here, $\ell^\perp$ is an $\mathrm{ad}_\ell(s)$-invariant subspace of ℓ, complementary to s.) Clearly, $z \subset c_\ell(s)$. Since

$$z \cap [z, c_\ell(s)] \subset z \cap [z, \ell] = \{0\},$$

we may choose a subspace W in $c_\ell(s)$ which is complementary to z and which contains $[z, c_\ell(s)]$. Then,

$$V = W + [s, \ell^\perp]$$

is an $ad_\ell(g^0)$-invariant subspace, which is complementary to $g^0 = s + z$ in ℓ. This implies that

$$\ell = V + g^0$$

is a quasi-gradation of ℓ.

The last claim follows from well known properties of the representations of reductive Lie algebras. □

The following example shows that, in general, a simple ideal of a semisimple quasi-graded Lie algebra of order 1 is not quasi-graded and that ℓ may not be decomposed into direct sum of quasi-graded simple Lie algebras. In other words, the analogue of statement 2 of Theorem 2.14 is not true for a generic non-flat quasi-gradation.

Example 3.2. Let $\ell = V + g^0$ be a simple quasi-graded Lie algebra of order 1, with g^0 effective. Let $\tilde{g}^0$ be a Lie algebra isomorphic to g^0 and fix a Lie isomorphism $\varphi : g^0 \to \tilde{g}^0$. Now, consider the semisimple Lie algebra

$$\ell' = \ell \oplus \tilde{g}^0.$$

It admits the quasi-gradation

$$\ell' = V' + g'^0$$

where

$$V' = \tilde{g}^0 + V,$$

$$g'^0 = \{X \mid X = Y + \varphi(Y),\ Y \in g^0\}.$$

Observe that, on the one hand, the reductive ideal $\tilde{g}^0$ of the quasi-graded Lie algebra ℓ' is a quasi-graded subalgebra of order 0. But, on the other hand, *the simple ideal* $\ell \simeq \ell'/\tilde{g}^0$ *of* ℓ' *is not a quasi-graded subalgebra of* ℓ*; and* ℓ' *cannot be decomposed into direct sum of quasi-graded ideals.*

Note that the quasi-graded Lie algebra $\ell = V + g^0$, from which we started, may be reconstructed from ℓ' by factorizing each subspace with respect to $\tilde{g}^0$:

$$\ell = \ell'/\tilde{g}^0 = V/\tilde{g}^0 + g'^0/\tilde{g}^0.$$

4. Reductive quasi-graded Lie algebras of order 2 with solvable isotropy subalgebra

Througout this and the next section, all Lie algebras are over the field of complex numbers $\mathbb{C}$.

Let H, E_+, E_- be a standard basis for the Lie algebra sl_2, i.e. such that

$$[H, E_\pm] = \pm 2E_\pm, \quad [E_+, E_-] = H.$$

This basis defines a gradation of sl_2 as follows:

$$sl_2 = V + g^0 + g^1 = \mathbb{C}E_- + \mathbb{C}H + \mathbb{C}E_+.$$

We call this gradation *the standard gradation of order 2 of sl_2*.

The following lemma is checked immediately.

Lemma 4.1. *Any quasi-gradation of order 2 of the Lie algebra sl_2 is flat and conjugate to the standard gradation.*

The remaining part of this section is devoted to the proof of the following proposition.

Proposition 4.2. *Let $\ell = V + g^0 + g^1$ be a quasi-graded reductive Lie algebra of order 2 and suppose that g^0 is solvable. Then, ℓ is isomorphic to the direct sum*

$$\ell = \ell' \oplus \ell'' = (sl_2 \oplus sl_2 \oplus \cdots \oplus sl_2) \oplus \ell'',$$

where $\ell' = sl_2 \oplus sl_2 \oplus \cdots \oplus sl_2$ is a graded subalgebra isomorphic to the direct sum of k copies of sl_2 with standard gradation, and ℓ'' is a quasi-graded subalgebra

$$\ell'' = V'' + {g^0}'' + {g^1}''$$

with ${g^1}''$ such that

$${g^1}'' = g^1 \cap z(\ell)$$

($z(\ell)$ = center of ℓ). If $g^0 + g^1$ is effective in ℓ, then ${g^1}'' = \{0\}$ and ℓ'' is of order at most 1.

Proof. Since $g^0 + g^1$ is a solvable subalgebra, it is contained in a Borel subalgebra b of ℓ. Let us fix a Cartan subalgebra $h \subset b$ and let us denote by R (resp. R^+) the system of roots (resp. positive roots) associated with h (resp. with h and b). A system of root vectors E_α, one per each $\alpha \in R^+$, span a nilpotent ideal n of b and we have:

$$b = h + n, \quad h \cap n = \{0\}.$$

Recall also that any element X of ℓ can be uniquely expressed as a linear combination of root vectors plus an element of the Cartan subalgebra.

We set

$$R_1 = \{\alpha \in R \mid \exists \alpha_1, \ldots, \alpha_m \in R,\ \lambda_1, \ldots, \lambda_m \in \mathbb{C},\ H \in h \text{ such that}$$
$$X = E_\alpha + \sum_{i=1}^{m} \lambda_i E_{\alpha_i} + H \in g^1\}.$$

Lemma 4.3.

1. $R_1 \subset R^+$;
2. $(R + R_1) \cap R \subset R^+$;

3. $(R^+ + R_1) \cap R = \emptyset$, *i.e., R_1 consists only of maximal roots;*
4. *If* $X = \sum_{i=1}^m \lambda_i E_{\alpha_i} + H \in g^1$, *then* $H \in z(\ell)$ *(= center of ℓ).*

Proof. 1. Note that, if $X = E_\alpha + \sum_{i=1}^m \lambda_i E_{\alpha_i} + H \in g^1 \subset b$, then the roots α and α_i belong to R^+, and so 1. is clear.

2. follows immediately from 1. and from the inclusion

$$[\ell, g^1] \subset g^0 + g^1 \subset b.$$

3. Suppose that, for some $\alpha \in R_1$, there exists a positive root $\beta > 0$ such that $\gamma = \alpha + \beta \in R$. Then, $-\gamma + \alpha = \beta < 0$. This contradicts 2. and proves 3.

4. Suppose that, for some $\beta \in R$, $[H, E_\beta] = \beta(H)E_\beta \neq 0$ and hence $0 \neq [X, E_\beta] \in g^0 + g^1 \subset b$. By considering, if necessary, the opposite root, we may assume that $\beta < 0$: but in this case, $[X, E_\beta] = \sum_{i=1}^m \lambda_i E_{\alpha_i + \beta} + \beta(H)E_\beta$ is not in the span of non-negative root vectors and this contradicts the definition of b. □

Corollary 4.4. *All roots from R_1 are mutually orthogonal and* $(R_1 \pm R_1) \cap R = \emptyset$.

Corollary 4.5. *The Lie algebra g^1 has the form*

$$g^1 = g(R_1) + z^1,$$

where $g(R_1)$ is the abelian subalgebra $g(R_1) = \langle E_\alpha,\ \alpha \in R_1 \rangle$, which is spanned by the root vectors E_α for all $\alpha \in R_1$, and z^1 is

$$z^1 = z(\ell) \cap g^1.$$

Proof. Let $X = \sum_{i=1}^m \lambda_i E_{\alpha_i} + Z \in g^1$, where $Z \in z(\ell)$, $\alpha_i \in R_1$ and all $\lambda_i \in \mathbb{C}$ are different from zero. It is sufficient to prove that $E_{\alpha_1} \in g^1$. We have

$$g^0 + g^1 \ni [E_{-\alpha_1}, X] = \lambda_1 [E_{-\alpha_1}, E_{\alpha_1}] \overset{\text{def}}{=} \lambda_1 H_{\alpha_1},$$

since $\alpha_i \pm \alpha_j \notin R$ for $i \neq j$, by Corollary 4.4. But then,

$$g^1 \ni [H_{\alpha_1}, X] = [H_{\alpha_1}, \sum_{i=1}^m \lambda_i E_{\alpha_i} + Z] = \lambda_1 \alpha_1(H_{\alpha_1}) E_{\alpha_1} = 2\lambda_1 E_{\alpha_1},$$

because $\alpha_1(H_{\alpha_j}) = 2\frac{<\alpha_1, \alpha_j>}{<\alpha_j, \alpha_j>} = 0$ for all $j \neq 1$. This concludes the proof. □

Lemma 4.6. *Any root $\alpha \in R_1$ is simple, that is it cannot be decomposed into the sum of two roots from R^+.*

Proof. Suppose that $\alpha = \beta + \gamma \in R_1$, for two positive roots $\beta > \gamma \in R^+$. The root vectors $E_{-\beta}$, $E_{-\gamma}$ may be written as

$$E_{-\beta} = u - X,\ E_{-\gamma} = v - Y,$$

where u, v are in V and X, Y are elements of $g^0 + g^1 \subset b$. Now, we calculate the commutator $[u, v]$:

$$[u, v] = [E_{-\beta} + X, E_{-\gamma} + Y] = cE_{-\alpha} + [E_{-\beta}, Y] + [X, E_{-\gamma}] + [X, Y] = cE_{-\alpha} + W$$

where $c \neq 0$ and W is a linear combination of root vectors E_δ, with the root δ such that $\delta > \min(-\beta, -\gamma) > -\alpha$.

By Proposition 2.12, we have that

$$[[u, v], E_\alpha] = 0.$$

On the other hand,

$$[[u, v], E_\alpha] = [cE_{-\alpha} + W, E_\alpha] = c[E_{-\alpha}, E_\alpha] + [W, E_\alpha] \neq 0,$$

since $0 \neq c[E_{-\alpha}, E_\alpha] \in h$ and $[W, E_\alpha]$ is a linear combination of root vectors associated with positive roots. This contradiction concludes the proof. □

Corollary 4.7. *The Lie algebra ℓ admits a decomposition into the direct sum $\ell = \ell' \oplus \ell''$ of two ideals, where $\ell' = ksl_2$ is the direct sum of k ideals, each of them isomorphic to sl_2, and ℓ'' is such that*

$$\ell'' \cap g^1 = z^1 \overset{\text{def}}{=} g^1 \cap z(\ell).$$

Proof. By Lemma 4.3 3., any root $\alpha \in R_1$ is maximal and defines a simple ideal of ℓ. By Lemma 4.6 this ideal is isomorphic to sl_2.

The last claim follows from Lemma 4.3 4. □

Lemma 4.8. *$g^0 \cap \ell' = h'$ is a Cartan subalgebra of ℓ'.*

Proof. Let $R_1 = \{\alpha_1, \dots, \alpha_k\}$. Then $g^1 = \langle E_{\alpha_i},\ i = 1, \dots, k\rangle$, and the vectors $H_{\alpha_i} = [E_{-\alpha_i}, E_{\alpha_i}] \in g^0 + g^1$ span a Cartan subalgebra of $\ell' \simeq ksl_2$. Now, we can write that

$$H_{\alpha_i} = H_i - X_i,\ H_i \in g^0,\ X_i \in g^1.$$

Then, using the fact that $[H_{\alpha_i}, g^1] \subset \mathbb{C}E_{\alpha_i}$, the commutator relations give

$$g^0 \ni [H_i, H_j] = [H_{\alpha_i} + X_i, H_{\alpha_j} + X_j] = [H_{\alpha_i}, X_j] + [X_i, H_{\alpha_j}] \in g^1.$$

So, $[H_i, H_j] = 0$ and, if we decompose X_j as $X_j = \sum_k \lambda_k{}^j E_{\alpha_k} + H'$ and X_i as $X_i = \sum_k \lambda_k{}^i E_{\alpha_k} + H''$, we also have that

$$0 = [H_{\alpha_i}, X_j] + [X_i, H_{\alpha_j}] = 2\lambda_i{}^j E_{\alpha_i} - 2\lambda_j{}^i E_{\alpha_j},$$

which implies $\lambda_i{}^j = \lambda_j{}^i = 0$ for all $i \neq j$. Hence, $X_j = \lambda_j E_{\alpha_j}$ and $H_j = H_{\alpha_j} + \lambda_j E_{\alpha_j}$, and this shows that H_j is an element of the j-th copy of sl_2. Moreover, being any element of the form $H_{\alpha_j} + \lambda_j E_{\alpha_j}$ semisimple in sl_2, we may claim that all vectors H_j span a Cartan subalgebra of $\ell' = ksl_2$. It is now easy to check that

$$h' = \langle H_j,\ j = 1, \ldots, k\rangle = g^0 \cap \ell'. \qquad \square$$

The proof of Proposition 4.2 is now an immediate consequence of the following:

Lemma 4.9. *The ideals ℓ' and ℓ'' are quasi-graded subalgebras of ℓ and, hence,*

$$\ell = \ell' \oplus \ell''$$

with

$$\ell' = V' + h' + g^{1'} = \ell' \cap V + \ell' \cap g^0 + \ell' \cap g^1,$$

$$\ell'' = V'' + h'' + g^{1''} = \ell'' \cap V + \ell'' \cap g^0 + \ell'' \cap g^1.$$

Moreover, we also have that $V' = \ell' \cap V = [h', V]$, $V'' = \ell'' \cap V = c_V(h')$ (= centralizer of h' in V), and $g^{1''} = \ell'' \cap g^1 = z(\ell) \cap g^1$.

Proof. Considering the adjoint representation of the Cartan subalgebra h' of the Lie subalgebra $\ell' \simeq ksl_2$ on $\ell = \ell' \oplus \ell''$ and the inclusion $h' \subset g^0$, we obtain

$$c_\ell(h') = h' \oplus \ell'' = c_V(h') + g^0 + g^1 \cap z(\ell),$$

$$V = [h', V] + c_V(h').$$

This shows that

$$\ell' = V' + h' + g^{1'},$$

where $V' = [h', V] = V \cap \ell'$; and

$$\ell'' = V'' + g^{0''} + g^{1''},$$

where $V'' = \ell'' \cap V = c_V(h')$ and $g^{0''} = g^0 \cap \ell''$. $\square$

5. The main theorem

We prove here:

Theorem 5.1. *Any reductive quasi-graded effective Lie algebra $\ell = V + g^0 + g^1$ of order 2 is the direct sum of two quasi-graded ideals*

$$\ell = \ell_1 \oplus \ell_2$$

where $\ell_1 = V_1 + g_1^0 + g^1$ is a flat semisimple quasi-graded Lie subalgebra of order 2 and $\ell_2 = V_2 + g_2^0$ (g_2^0 possibly equal to $\{0\}$) is a reductive quasi-graded Lie subalgebra of order less or equal to 1. In particular, $g^0 = g_1^0 \oplus g_2^0$ and $V = V_1 + V_2$.

The proof follows from several lemmata.

In the following, by *maximal* ad_ℓ*-diagonalizable subalgebra*, we mean a maximal subalgebra in the class of all abelian subalgebras, such that the adjoint actions of their elements on ℓ are semisimple.

Lemma 5.2. *There exists a non-trivial maximal* ad_ℓ*-diagonalizable subalgebra* h_0 *of* g^0 *such that its centralizer* $c = c_\ell(h_0)$ *has the form*

$$c = c_V + c_0$$

where c_V *is* $c \cap V$ *and* $c_0 = c \cap g^0$ *is a solvable subalgebra of* g^0.

Proof. Let $g^0 = s + r$ be a Levi–Malcev decomposition of g^0 into the sum of the radical r and a semisimple subalgebra s.

In the case $s \neq \{0\}$, denote by h_0 an (abelian) maximal ad_ℓ- diagonalizable subalgebra of g^0 that contains a Cartan subalgebra h_s of s. If $H \in h_0$ and if we express H as a sum $H = S + R$, where $S \in s$ and $R \in r$, then we have that

$$[h_s, S] = \{0\}.$$

Hence, $S \in h_s$ and $R \in h_0$. In other words, $h_0 = h_s + h_r$, with $h_r = h_0 \cap r$.

In the case $s = \{0\}$, by Proposition 4.2 and effectiveness, we know that $\ell = \ell' \oplus \ell''$, where ℓ' is a flat quasi-graded ideal of order 2, isomorphic to k copies of sl_2, while ℓ'' is a quasi-graded ideal of order at most 1. Any maximal ad_ℓ-diagonalizable subalgebra must be of the form $h_0 = h_s + h_r = h_0 \cap \ell' + h_0 \cap \ell''$. By Lemma 4.8, we may assume that $h_s = g^0 \cap \ell'$.

Since $h_0 \subset g^0$, the centralizer $c = c_\ell(h_0)$ is a quasi-graded subalgebra

$$c = V \cap c + g^0 \cap c + g^1 \cap c = c_V + c_0 + c_1.$$

Now, we may again use the structural result, given by Proposition 4.2, for what concerns the Lie algebra $c = c_V + c_0 + c_1$: in other words, we have that $c = c' + c''$ with $c' \simeq ksl_2$ and c'' such that $c'' \cap c^1 \subset z(c)$.

Recall that $h_0 \subset c_0 = c_0 \cap c' + c_0 \cap c''$. Moreover, any element $X \in c_0 \cap c'$ must be in h^0, because $c_0 \cap c'$ is a Cartan subalgebra of ksl_2 and hence is ad_ℓ-diagonalizable.

Since c_1 commutes with h_0 (and hence with $c' \cap c^0$), by Proposition 4.2, $c_1 \subset c'' \cap g^1 \subset z(c)$ ($z(c)$ = center of c). Then, c_1 commutes also with any fixed Cartan subalgebra h of ℓ that contains h_0. Therefore

$$c_1 \subset c_\ell(h) = h$$

($c_\ell(h)$ = centralizer of h in ℓ); but $c_1 \subset g^1$ is made only of ad_ℓ-nilpotent elements: this implies that $c_1 = \{0\}$. □

Corollary 5.3. *There exists a Borel subalgebra* b *of* ℓ*, which contains* $c_0 + g^1$ *(*c_0 *= centralizer of* h_0 *in* g^0*) and a Cartan subalgebra* $h \subset b$ *of* ℓ*, which contains the non-trivial maximal* ad_ℓ*-diagonalizable subalgebra* h_0 *of* g^0 *of Lemma* 5.2. *Moreover,* g^1 *is contained in* $n = [b, b]$.

Proof. Note that $c_0 + g^1$ is a solvable subalgebra of ℓ. Hence it is contained in a

Borel subalgebra b. Also, there exists a Cartan subalgebra h of b which contains $h_0 \subset c_0$. At last, using Lemma 5.2, we have that

$$g^1 = [h_0, g^1] \subset [b, b] = n. \qquad \Box$$

Now, in all that follows, we fix a Borel subalgebra b and a Cartan subalgebra $h \subset b$ as in Lemma 5.2 and Corollary 5.3. We denote by R (resp. by R^+) the system of roots (resp. positive roots) associated with h and b. Recall that $b = h + n$, where the subalgebra $n = [b, b]$ is spanned by the root vectors E_α, $\alpha \in R^+$.

We also set

$$R_1 = \{\alpha \in R, \quad \text{such that } \exists X = E_\alpha + \sum_{i=1}^{k} \lambda_i E_{\alpha_i} \in g^1, \quad \text{for some } 0 \neq \lambda_i \in \mathbb{C}, \alpha_i \in R\}.$$

By Corollary 5.3, we know that $R_1 \subset R^+$.

Finally, we also fix a system of root vectors E_α, which form a Chevalley system for ℓ (cfr. [Bo]). As before, we set $H_\alpha = [E_{-\alpha}, E_\alpha]$, for any root α, and we will denote by $<,>$ the standard scalar product on the real vector space generated by the roots.

Recall that the following relation holds:

$$\beta(H_\alpha) = 2\frac{<\alpha, \beta>}{<\alpha, \alpha>}$$

for any pair of roots α and β.

We now continue with a small technical lemma.

Lemma 5.4. *If two roots $\alpha_1, \alpha_2 \in R$ are such that*

$$\alpha_1(H_{\alpha_2}) = \alpha_2(H_{\alpha_1}) = 2,$$

then $\alpha_1 = \alpha_2$.

Proof. We have that

$$\alpha_1(H_{\alpha_2}) = 2\frac{<\alpha_1, \alpha_2>}{<\alpha_2, \alpha_2>} = 2\cos(\theta_{12})\frac{\|\alpha_1\|}{\|\alpha_2\|}.$$

So, the hypothesis implies that $\frac{\|\alpha_1\|}{\|\alpha_2\|} = \frac{\|\alpha_2\|}{\|\alpha_1\|} = 1$ and that $\cos(\theta_{12}) = 1$, i. e. that the angle between the two roots is 0. $\Box$

Lemma 5.5. *Let $X = \sum_1^k \lambda_i E_{\alpha_i} \in g^1$ (with all $\lambda_i \neq 0$) and suppose it is a weight vector of the previously defined subalgebra h_0, corresponding to the weight $\bar{\alpha} = \alpha_i|_{h_0}$, $\forall i = 1, \dots, k$. Then, for all $i, j = 1, \dots k$,*

$$[E_{-\alpha_i}, E_{\alpha_j}] = 0 \quad \text{if } i \neq j \tag{5.1}$$

and

$$[E_{-\alpha_i}, E_{\alpha_i}] = H_{\alpha_i} \in g^0. \tag{5.2}$$

Proof. Suppose not. Then, for some $i \neq j$, we have that $\alpha_i - \alpha_j \in R^+$ and hence we may consider the following decomposition into root vectors:

$$[E_{-\alpha_i}, X] = \lambda_j E_{-\alpha_i+\alpha_j} + \sum_{k \neq j} \lambda_j [E_{-\alpha_i}, E_{\alpha_k}],$$

where, by hypothesis, $-\alpha_i + \alpha_j$ is a negative root.

On the other hand, the vector

$$[E_{-\alpha_i}, X] \in c_0 \subset b$$

and its decomposition into sum of root vectors consists only of vectors corresponding to non-negative roots. This contradiction proves (5.1).

From (5.1), it follows immediately that $g^0 + g^1 \ni [E_{-\alpha_i}, X] = \lambda_i [E_{-\alpha_i}, E_{\alpha_i}] = \lambda_i H_{\alpha_i}$: hence, $H_{\alpha_i} \in (g^0 + g^1) \cap h = h_0 \subset g^0$. □

Lemma 5.6. *Any weight vector for h_0 in g^1 is also a root vector for h. As a consequence, the space g^1 is spanned by root vectors, i.e.*

$$g^1 = g(R_1) \stackrel{\text{def}}{=} \langle E_\alpha, \ \alpha \in R_1 \rangle.$$

Proof. Suppose not. Then there exists a weight vector $0 \neq X = \sum_1^k \lambda_i E_{\alpha_i} \in g^1$, with all $\lambda_i \neq 0$ and $k > 1$, such that, for all $i = 1, \ldots, k$,

$$E_{\alpha_i} \notin g^1.$$

We shall assume that $k > 1$ is the minimal integer for all vectors of such a kind. By Lemma 5.5, we also know that

$$H_{\alpha_i} = [E_{-\alpha_i}, E_{\alpha_i}] \in g^0.$$

So,

$$[H_{\alpha_i}, X] = \bar{\alpha}(H_{\alpha_i}) X = \sum_{j=1}^{k} \lambda_j \alpha_j (H_{\alpha_i}) E_{\alpha_j}.$$

This implies that, for all $i, j = 1, \ldots, k$,

$$\alpha_j(H_{\alpha_i}) = \alpha_i(H_{\alpha_i}) = 2 = \alpha_j(H_{\alpha_j}) = \alpha_i(H_{\alpha_j})$$

and, by Lemma 5.4, this means that $\alpha_i = \alpha_j$. Contradiction. □

Corollary 5.7. *Any weight space for h_0 in g^1 is 1-dimensional and it is generated by a root vector.*

Proof. Suppose that the weight space corresponding to a weight $\bar{\alpha}$ is not 1-dimensional. By Lemma 5.6, there exists at least two root vectors E_{α_1} and E_{α_2} in g^1 such that

$$\alpha_1|_{h_0} = \alpha_2|_{h_0} = \bar{\alpha}.$$

Therefore,

$$[H_{\alpha_1}, E_{\alpha_1} + E_{\alpha_2}] = \bar{\alpha}(H_{\alpha_1})(E_{\alpha_1} + E_{\alpha_2}) = 2E_{\alpha_1} + \alpha_2(H_{\alpha_1})E_{\alpha_1}.$$

Hence, $2 = \alpha_2(H_{\alpha_1})$ and, with a symmetric argument, $2 = \alpha_1(H_{\alpha_2})$. By Lemma 5.4, this implies that $\alpha_1 = \alpha_2$. Contradiction. □

Lemma 5.8. *Let us denote by*

$$g(-R_1) = \langle E_{-\alpha}, \ \alpha \in R_1 \rangle$$

the subspace that is spanned by the root vectors $E_{-\alpha}$, with $\alpha \in R_1$. Then, $g(-R_1)$ is an abelian subalgebra and it is contained in V.

Proof. It is sufficient to prove that $g(-R_1) \subset V$. Let us decompose a vector $E_{-\alpha}$ into a sum of the form

$$E_{-\alpha} = v + X + Y,$$

where $v \in V$, $X \in g^0$ and $Y \in g^1$. Let us first prove that $X = 0$.

Write X in the following form:

$$X = H + \sum \lambda_j E_{\beta_j}$$

where $H \in h_0$ and all λ_j are $\neq 0$. Since $H_\alpha = [E_{-\alpha}, E_\alpha] \in g^0$ and $[H_\alpha, E_{-\alpha}] = -2E_{-\alpha}$, we have that

$$[H_\alpha, X] = -2X,$$

and therefore,

$$H = 0, \quad [H_\alpha, \lambda_j E_{\beta_j}] = -2\lambda_j E_{\beta_j}.$$

Hence, for all j, we also have that

$$\beta_j(H_\alpha) = -2.$$

Moreover, using again the fact that $[E_{-\alpha}, E_\alpha] = [v + X + Y, E_\alpha] = H_\alpha \in g^0$, we have that

$$g^1 \ni [X, E_\alpha] = 0$$

because of the quasi-gradation properties. Therefore, for all j, $[E_\alpha, \lambda_j E_{\beta_j}] = 0$. This shows that the α-sequence of roots, generated by the root β_j, have length 3, with the highest root equal to β_j. In other words, it is of the form (cfr. [Bo])

$$\{\beta_j - 2\alpha, \beta_j - \alpha, \beta_j\}.$$

Note that

$$0 \neq E_{\beta_j} = k[[E_{\beta_j - 2\alpha_i}, E_\alpha], E_\alpha] \in [[\ell, g^1], g^1] \subset g^1.$$

Hence,

$$X = \sum \lambda_j E_{\beta_j} \in g^1 \cap g^0 = \{0\}.$$

It remains to show that $Y = 0$.

Suppose not. Then Y is a weight vector in g^1 for h_0 (it corresponds to the weight $-\bar{\alpha} = -\alpha|_{h_0}$): for Corollary 5.7, it coincides with a root vector, say λE_β.

Consider now the vector $E_{-\beta}$: by the previous arguments we way write it as

$$E_{-\beta} = v' + Y'$$

with $v' \in V$ and $Y' \in g^1$. Also Y' is a weight vector and it corresponds to the weight $-(-\bar{\alpha}) = \bar{\alpha} = \alpha|_{h_0}$. So, by Corollary 5.7, we have that $Y' = \mu E_\alpha$. Consider the commutator

$$[H_\beta, E_{-\alpha}] = -\alpha(H_\beta)E_{-\alpha} = -\alpha(H_\beta)v + 2\lambda E_\beta.$$

So, $\alpha(H_\beta) = -2$ and, by considering the commutator $[H_\alpha, E_{-\beta}]$, we also have that $\beta(H_\alpha) = -2$. Then, by Lemma 5.4, this implies that $\alpha = -\beta$ and that $E_{-\alpha} \in g^1$. But this is impossible, proving that $Y = 0$. □

Denote by ℓ_1 the Lie subalgebra of ℓ, which is generated by $g^1 = g(R_1)$ and $g(-R_1)$. It is clear that ℓ_1 is a semisimple regular subalgebra of ℓ. Moreover,

Lemma 5.9. *ℓ_1 is a flat quasi-graded subalgebra, with the gradation given by*

$$\ell_1 = V_1 + {g^0}_1 + g^1,$$

where $V_1 \stackrel{\text{def}}{=} g(-R_1)$ and ${g^0}_1$ is a reductive subalgebra of g^0, which is spanned by the vectors

$$H_\alpha = [E_{-\alpha}, E_\alpha] \quad \text{for } \alpha \in R_1 \quad \text{and} \quad E_{\alpha-\beta} \quad \text{for } \alpha - \beta \in (R_1 - R_1) \cap R.$$

Proof. It is sufficient to prove that

$${g^0}_1 = [g(-R_1), g(R_1)] = \langle [E_{-\beta}, E_\alpha],\ \alpha, \beta \in R_1 \rangle$$

is a subalgebra of g^0. In fact, from this it will follow also that

$$[{g^0}_1, g(R_1)] \subset g(R_1) = g^1$$

and, by duality, $[{g^0}_1, g(-R_1)] \subset g(-R_1)$.

The claim follows simply from the Jacobi identities. In fact, let $[E_\alpha, E_{-\beta}]$ and $[E_{\alpha'}, E_{-\beta'}]$ be two elements of ${g^0}_1$. Then, the Lie brackets between them are of the form

$$[[E_\alpha, E_{-\beta}], [E_{\alpha'}, E_{-\beta'}]] =$$

$$[E_\alpha, [E_{-\beta}, [E_{\alpha'}, E_{-\beta'}]]] - [E_{-\beta}, [E_\alpha, [E_{\alpha'}, E_{-\beta'}]]]. \tag{5.3}$$

Note that $[E_\alpha, [E_{\alpha'}, E_{-\beta'}]]$ is in g^1 and that it is proportional to some E_γ; similarly $[E_{-\beta}, [E_{\alpha'}, E_{-\beta'}]]$ is in $g(-R_1)$, because, if it is not trivial, it is dual to $[E_\beta, [E_{-\alpha'}, E_{\beta'}]] \in g^1$. Hence the Lie brackets (5.3) can be always expressed as sum of elements of the form $[E_\alpha, E_{-\gamma}]$ or $[E_{-\beta}, E_\gamma]$. □

Lemma 5.10.

1. *${g^0}_1$ is orthogonal (w.r.t. the Cartan–Killing form $\mathcal{B}$) to the centralizer $c_{g^0}(g^1)$ of g^1 in g^0;*
2. *$c_{g^0}(g^1)$ is an ideal of g^0.*

Proof. 1. Let X be an element of $({g^0}_1)^\perp$, i.e. the subspace of g^0 of elements, which are orthogonal to ${g^0}_1$. Note that, being ${g^0}_1$ reductive, $\mathcal{B}|_{{g^0}_1}$ is not degenerate and hence $({g^0}_1)^\perp \cap {g^0}_1 = \{0\}$ and $g^0 = ({g^0}_1)^\perp + {g^0}_1$. Suppose that $X \notin c_{g^0}(g^1)$, and hence that, for some $E_\alpha \in g^1$,

$$[X, E_\alpha] \neq 0.$$

Then, there exists some $E_{-\beta} \in g^{-1}$ such that

$$\mathcal{B}([X, E_\alpha], E_{-\beta}) \neq 0.$$

This cannot be because

$$\mathcal{B}([X, E_\alpha], E_{-\beta}) = -\mathcal{B}(X, [E_{-\beta}, E_\alpha]) = 0,$$

since $[E_{-\beta}, E_\alpha] \in {g^0}_1$. Similarly, if $X \in c_{g^0}(g^1)$,

$$\mathcal{B}(X, [E_{-\beta}, E_\alpha]) = -\mathcal{B}([X, E_\alpha], E_{-\beta}) = 0,$$

for all $[E_{-\beta}, E_\alpha] \in {g^0}_1$ and, therefore, $X \in ({g^0}_1)^\perp$.

2. follows from the fact that $c_{g^0}(g^1)$ is the kernel of the adjoint representation of g^0 on g^1. □

The last steps of the proof of Theorem 5.1 consist in showing that ℓ splits into the direct sum of ℓ_1 and an ideal which is orthogonal to ℓ_1. We will prove that such a complement to ℓ_1 is given by

$$V_2 + {g^0}_2,$$

where ${g^0}_2 \stackrel{\text{def}}{=} c_{g^0}(g^1)$ and V_2 is defined as

$$V_2 = \{v \in V \mid \mathcal{B}(v, E_\alpha) = 0 \ \forall E_\alpha \in g^1\}$$

Since g^1 is dual to $V_1 = g(-R_1)$, it is simple to check that $V = V_2 + V_1$ and that $V_2 \cap V_1 = \{0\}$.

Lemma 5.11.

1. *$[V, V_1] = 0$;*
2. *$V_2 + {g^0}_2$ is a subalgebra ℓ_2 of ℓ.*

Proof. 1. Suppose that, for some $E_{-\alpha} \in V_1$, $v \in V$,

$$[v, E_{-\alpha}] \neq 0.$$

Then, decomposing the vectors v and $[v, E_{-\alpha}]$ into sums of root vectors, we can check that $[[v, E_{-\alpha}], E_\alpha] \neq 0$. But this cannot be by Proposition 2.12.

2. For any $E_\alpha \in g^1$, $v \in V$, $A \in {g^0}_2$,

$$\mathcal{B}([v, A], E_\alpha) = \mathcal{B}(v, [A, E_\alpha]) = 0.$$

Therefore $[V, {g^0}_2] \subset V_2$, because otherwise it would not be orthogonal to g^1.

For any $v, v' \in V_2$ and $E_\alpha \in g^1$, we have that $H_\alpha = [E_{-\alpha}, E_\alpha] \in {g^0}_1$ and that

$$\begin{aligned}\mathcal{B}([v, v'], E_\alpha) &= \frac{1}{2}\mathcal{B}([v, v'], [H_\alpha, E_\alpha]) \\ &= -\frac{1}{2}\mathcal{B}([[v, v'], E_\alpha], H_\alpha) = 0\end{aligned}$$

by Proposition 2.12. Furthermore,

$$\begin{aligned}\mathcal{B}([v, v'], E_{-\alpha}) &= -\frac{1}{2}\mathcal{B}([v, v'], [H_\alpha, E_{-\alpha}]) \\ &= \frac{1}{2}\mathcal{B}([[v, v'], E_{-\alpha}], H_\alpha) \\ &= \frac{1}{2}\mathcal{B}([[v, E_{-\alpha}], v'], H_\alpha) + \frac{1}{2}\mathcal{B}([v, [v', E_{-\alpha}]], H_\alpha) = 0\end{aligned}$$

by 1. This means that $[V_2, V_2]$ is orthogonal to $V_1 + g^1$ and therefore that $[V_2, V_2] \subset V_2 + g^0$. On the other hand, $[V_2, V_2]$ commutes with g^1 and hence,

$$[V_2, V_2] \subset V_2 + {g^0}_2.$$

Since we already know that $[{g^0}_2, {g^0}_2] \subset {g^0}_2$, this concludes the proof. □

Lemma 5.12. *$\ell_2 = V_2 + {g^0}_2$ is an ideal of ℓ.*

Proof. For any $E_\alpha \in g^1$, $E_{-\beta} \in V_1$ and $A \in {g^0}_2$

$$\mathcal{B}([E_{-\beta}, A], E_\alpha) = -\mathcal{B}(E_{-\beta}, [E_\alpha, A]) = 0.$$

Hence, $[V_1, {g^0}_2] \subset V_2$. Moreover $[V_1, V_2] = 0$ by Lemma 5.11 1.

Then, consider $v \in V_2$, $[E_\alpha, E_{-\beta}] \in {g^0}_1$ and $E_\gamma \in g^1$:

$$\mathcal{B}([v, E_\gamma], [E_\alpha, E_{-\beta}]) = \mathcal{B}(v, [E_\gamma, [E_\alpha, E_{-\beta}]]) = 0,$$

because $[E_\gamma, [E_\alpha, E_{-\beta}]] \in g^1$. Therefore,

$$[V_2, g^1] \subset {g^0}_2 \quad (= ({g^0}_1)^\perp)$$

and

$$[V_1 + g^1, V_2 + {g^0}_2] \subset V_2 + {g^0}_2.$$

This implies also that $[{g^0}_1, V_2 + {g^0}_2] \subset V_2 + {g^0}_2$ and concludes the proof. □

Proof of Theorem 5.1. From the construction of ℓ_1 and of $V_2 + g^0{}_2$, it is clear that

$$\ell_1 = [\ell_1, [\ell_1, g^1]]$$

and that

$$\mathcal{B}(V_2 + g^0{}_2, g^1) = 0$$

Therefore

$$\mathcal{B}(V_2 + g^0{}_2, \ell_1) = \mathcal{B}(V_2 + g^0{}_2, [\ell_1, [\ell_1, g^1]])$$
$$\subset \mathcal{B}([\ell_1, [\ell_1, V_2 + g^0{}_2]], g^1) = \mathcal{B}(V_2 + g^0{}_2, g^1) = 0.$$

Therefore, ℓ_1 coincides with the ideal that is orthogonal to $V_2 + g^0{}_2$; it is also clear that there is no center in ℓ_1 and that it is a semisimple flat, quasi-graded Lie algebra subalgebra; finally, we also have that $\ell_2 = V_2 + g^0{}_2$ is a quasi-graded reductive Lie subalgebra of order less or equal to 1 (according to $g^0{}_2$ equal or not equal to $\{0\}$). This concludes the proof. □

References

[Al] D. V. Alekseevsky, Maximally homogeneous G-structures and filtered Lie algebras, Dokl. Akad. Nauk SSSR 299 (3) (1988), 521–525, (in Russian); English transl. in Soviet Math. Dokl. 37 (1988), 381–384.

[Bo] N. Bourbaki, Éléments de Mathématiques - Groupes et Algèbres de Lie, Ch.4, 5 et 6, Masson, Paris 1981.

[Gu] V. Guillemin, A Jordan–Hölder decomposition for a certain class of infinite dimensional Lie algebras, J. Differential Geom. 2 (1968), 313–345.

[Ja] N. Jacobson, Lie algebras, Dover Publication, Inc., New York 1979.

[Ka] I. L. Kantor, Trudy Sem. Vektor Tenzor Anal. 3 (1966), 310–398 (in Russian).

[KN] S. Kobayashi and T. Nagano, On filtered Lie algebras and geometric structures I, II, J. Math. Mech. 13/14 (1964/1965), 875–907/513–521.

[LT] F. Lastaria and F. Tricerri, Curvature orbits and locally homogeneous Riemannian manifolds, Ann. Mat. Pura Appl. 165 (1993), 121–131.

[Oc] T. Ochiai, Geometry associated with semisimple flat homogeneous spaces, Trans. Amer. Math. Soc. 152 (1970), 159–193.

[Sp] A. Spiro, Lie pseudogroups and locally homogeneous Riemannian spaces, Boll. U.M.I. 6-B (1992), 843–872.

[Ta] N. Tanaka, On the equivalence problems associated with a certain class of homogeneous spaces, J. Math. Soc. Japan 17 (1965), 103–139.

A problem on an exceptional domain

Walter L. Baily, Jr.

We wish to consider a problem of an algebraico-geometric nature connected with certain Hermitian symmetric tube domains, and in particular with the 27-dimensional tube domain $\mathcal{T}_e$ on which a certain real form of E_7 acts.

First some notation. Let $V \cong \mathbb{R}^n$ be an n-dimensional real vector space and $\mathcal{K}$ be a self-adjoint homogeneous convex cone such that the tube domain $\mathcal{T} = V + i\mathcal{K}$ is a complex Hermitian symmetric domain whose group G of holomorphic automorphisms is a semi-simple real algebraic Lie group defined over $\mathbb{Q}$. Let $\Gamma \hookrightarrow G(\mathbb{Q})$ be an arithmetic subgroup of $G(\mathbb{Q})$.

We have: $\mathcal{T}_e = \mathcal{J}_e + i\mathcal{P}_e$, where $\mathcal{J}_e$ is the 27-dimensional exceptional real Jordan algebra of 3×3 Hermitian matrices over the Cayley division algebra C and $\mathcal{P}_e$ is its cone of positive elements. There is a certain "nice" unicuspidal arithmetic subgroup Γ_e of $\mathrm{Hol}(\mathcal{T}_e)$ for which the associated Eisenstein series have rational Fourier coefficients [Ba1].

Now if $\mathcal{T} = \mathcal{H}_n$ is the Siegel upper half-space and $\Gamma = \Gamma_n$, the Siegel modular group of degree n, then the orbits of Γ_n in $\mathcal{H}_n$ correspond one-to-one to the isomorphism classes of principally polarized Abelian varieties (p.p.A.v's) of dimension n. Moreover, there is a Γ_n-invariant complex analytic closed subset $\mathcal{J}_n$ of $\mathcal{H}_n$ of which a Zariski-open subset $\mathcal{J}_n^{\#}$ is in one-to-one correspondence with canonically polarized Jacobian varieties of curves of genus n, hence, via Torelli's theorem, with the isomorphism classes of non-singular curves of genus n. If $n = 3$, then $\mathcal{J}_3 = \mathcal{H}_3$, and this special case is important for the considerations which follow.

However, there is no known interpretation of the space of orbits of Γ_e in $\mathcal{T}_e$ as the space of moduli of some family of polarized algebraic varieties. The problem is to seek such a family.

We approach this problem by considering the four Severi varieties of Zak [LaZ], S_n, $n = 1, 2, 3, 4$, where $\dim(S_n) = 2^n$. These are given explicitly as follows ($\mathbb{P}^n$ is the projective space of dimension n):

$$\begin{aligned} S_1 &= \mathbb{P}^2 \hookrightarrow \mathbb{P}^5 \ \text{(Veronese imbedding)};\\ S_2 &= \mathbb{P}^2 \times \mathbb{P}^2 \hookrightarrow \mathbb{P}^8 \ \text{(Segre imbedding)};\\ S_3 &= G(2,6) \hookrightarrow \mathbb{P}^{14} \ \text{(Plucker imbedding)},\\ S_4 &= C\mathbb{P}^2 \hookrightarrow \mathbb{P}^{26}, \end{aligned}$$

where $G(2,6)$ is the Grassmannian of planes in 6-space, C stands for the Cayley numbers, and $C\mathbb{P}^2$ is the Cayley projective plane, realized as the projective variety of the primitive idempotents in the exceptional 27-dimensional Jordan algebra $\mathcal{J}$ of three-by-three Hermitian matrices over $\mathcal{C}$, whose projective space is $\mathbb{P}^{26}$, viewed as a 27-dimensional irreducible module of E_6.

Now observe that the *generic* quadric hypersurface section $Q \cap S_2$ of S_2 as imbedded in $\mathbb{P}^5$, where Q is a quadric in $\mathbb{P}^5$, is a non-singular, non-hyperelliptic plane quartic curve C of genus 3 in $\mathbb{P}^2$. Then, as mentioned above, the moduli of such curves are essentially given, via Torelli's theorem, by the orbits of Γ_3 in $\mathcal{H}_3$. By "essentially" we mean:

three-by-three complex unitary matrices.

$D_2 = \mathbb{H}_3$, the 9-dimensional tube domain in $\mathbb{C}^9 \cong \mathbb{M}_3(\mathbb{C})$, the three-by-three complex matrices $Z = X + iY$, where X and Y are complex Hermitian three-by-three matrices, and $\mathbb{H}_3$ consists of those for which Y is positive definite, and can be written as $U(3,3)/K_2$, where $K_2 = U(3) \times U(3)$.

$D_3 = \mathbb{Q}_3$ is the tube domain in $\mathbb{C}^{15}$ analogous to $\mathbb{H}_3$ with quaternion Hermitian 3×3 matrices in place of complex Hermitian matrices, and $\mathbb{Q}_3$ can be written as $SO^*(12)/K_3$, with $K_3 = U(6)$. (The notation $SO^*(2n)$ is from Helgason: "Differential Geometry and Symmetric Spaces", first ed., p. 354.).

$D_4 = \mathcal{T}_e$ is the tube domain $\{Z = X + iY \mid X, Y \in$ 27- dimensional real exceptional Jordan algebra $\mathcal{J}$ of 3×3 Hermitian matrices over $\mathcal{C}$ such that Y is positive definite (written $Y > 0\}$. We may write $D_4 = E_{7(-25)}/K_4$, where $K_4 = E_{6(-78)} \times C'$, C' being the unit circle.

We note in each of the above cases that *there is a natural action of K_n on the ambient manifold of S_n*, if we identify that ambient manifold with the projective space of the complexification of the appropriate Jordan algebra of Hermitian matrices over R, C, H or $\mathcal{C}$; thus,

$$\begin{aligned}
\mathbb{P}^5 &= \mathbb{P}\Sigma_{3\mathbb{C}}),\ \Sigma_3 \text{ being the } 3\times 3 \text{ symmetric real matrices;}\\
\mathbb{P}^8 &= \mathbb{P}(H_{3\mathbb{C}}),\ H_3 \text{ being the complex Hermitian } 3\times 3 \text{ matrices;}\\
\mathbb{P}^{14} &= \mathbb{P}(Q_{3\mathbb{C}}),\ Q_3 \text{ being the quaternion Hermitian } 3\times 3 \text{ matrices;}\\
\mathbb{P}^{26} &= \mathbb{P}(\mathcal{J}_{3\mathbb{C}}),\ \mathcal{J}_3 \text{ being the } 3\times 3 \text{ Cayley Hermitian matrices.}
\end{aligned}$$

Motivated by these considerations, we try the simplest first step. Namely, we examine the configurations of algebraic varieties which arise when for a generic quadric hypersurface $Q \subset \mathbb{P}^8$ we construct the non-singular 3-fold

$$F = F_Q = Q \cap (\mathbb{P}^2 \times \mathbb{P}^2) \subset \mathbb{P}^8.$$

A hyperplane section of F is a K-3 surface in two different ways: On the one hand the generic hyperplane section $F^H = F \cap H$ is a canonical curve of genus 7 in $\mathbb{P}^6$. On the other hand, F is fibered into conics over $\mathbb{P}^2$ with a sextic branch curve. We now explain this more in detail. Explicitly, F is fibered into conics over $\mathbb{P}^2$ as follows:

$$\pi = \pi_Q : Q \cap (\mathbb{P}^2 \times \mathbb{P}^2) \longrightarrow \mathbb{P}^2,$$

where π_Q is the restriction of pr_2 to $F = F_Q$. For $s \in \mathbb{P}^2$, $\pi^{-1}(s) = Q \cap (\mathbb{P}^2 \times \{s\})$ which is a plane conic in the coordinates of the first factor by virtue of the nature of the Segre imbedding $\mathbb{P}^2 \times \mathbb{P}^2 \longrightarrow \mathbb{P}^8$. Let Δ_π be the locus of $s \in \mathbb{P}^2$ such that

$\pi^{-1}(s) = F_s$ is the union of two lines. Let $r = [r_0 : r_1 : r_2]$ resp. $s = [s_0 : s_1 : s_2]$ be the coordinates of the first resp. second factor $\mathbb{P}^2$, and $t_{ij}, i, j = 0, 1, 2$, be the coordinates in $\mathbb{P}^8$. Suppose the quadric hypersurface Q in $\mathbb{P}^8$ is given by $A(t) = 0$, where A is the quadratic form

$$A(t) = \sum a_{ijkl} t_{ij} t_{kl},$$

so that with $t_{ij} = r_i s_j$ we have

$$F_s : \sum b_{ik} r_i r_k = 0,$$

with $b_{ik} = \sum_{j,l} a_{ijkl} s_j s_l$. The conic degenerates to two lines if and only if the discriminant $|b_{ik}| = \det(b_{ik}) = 0$, and $|b_{ik}|$ is a homogeneous cubic polynomial in $\{b_{ik}\}$, hence (for fixed a_{ijkl}) is a homogeneous sextic polynomial in s_0, s_1, s_2. Therefore, Δ_π is a sextic plane curve in $\mathbb{P}^2$. It then follows from known formulae [Is, sec. 14.5] that the third Betti number $b_3(F)$ is equal to 18. "In general", Δ_π is non-singular (for generic choice of a_{ijkl}. Therefore, by [B, Théoreme 2.1 (with $n = 1$)] the level 3 Hodge structure of F is of the form

$$H^3(F) = H^{2,1}(F) \oplus H^{1,2}(F),$$

so that the intermediate Jacobian $J(F) = H^{2,1}(F)/H^3(F, \mathbb{Z})$ is a normally polarized Abelian variety of dimension $(1/2)b_3(F) = 9$.

In fact, the family of varieties:

$$\{ F_Q \mid Q \text{ a quadric hypersurface in } \mathbb{P}^8 \}$$

does not seem as well suited to the investigation of moduli problems as the family

$$\{ F^H = F_Q \cap H \mid Q \text{ a quadric hypersurface and } H\text{, a hyperplane in } \mathbb{P}^8 \}.$$

For example, there is no apparent natural relationship between F_Q and any particular one of its hyperplane sections F^H, whereas the latter seem more naturally related to the moduli problems we wish to study, as we shall see.

So let $Q' = Q \cap H$ be a generic irreducible quadric variety of codimension 2 in $\mathbb{P}^8$. As before, $S_2 = \mathbb{P}^2 \times \mathbb{P}^2 \hookrightarrow \mathbb{P}^8$ is a Severi variety and we let

$$F' = F_{Q'} = Q' \cap S_2 \subset \mathbb{P}^8$$

and define

$$\pi' = \pi_{Q'} : F' \longrightarrow \mathbb{P}^2,$$

where π' is the restriction of pr_2 to F'. This exhibits F' as a double covering of $\mathbb{P}^2$, with a branch curve C which is the zero-locus of the discriminant Δ of the quadratic equation whose roots are coordinates of the two points of F' over a given point s of $\mathbb{P}^2$. Direct calculation shows that Δ is homogeneous of degree two in both the coefficients of Q and in the coefficients of H, after factoring out a quadratic polynomial factor which is nowhere vanishing on a given Zariski-open subset of $\mathbb{P}^2$. Hence, Δ is homogeneous of degree 6 in the homogeneous coordinates of s, and therefore the branch curve C is a sextic curve. Generically C is non-singular and F' is a K-3-surface.

We know [V: §§ 2.7, 2.8; B: § 6.23; Di] the following:

Proposition 1. *A sufficiently general homogeneous sextic polynomial* $\Sigma(s) = \Sigma(s_0, s_1, s_2)$ *can be expressed as a symmetric determinant*

$$\det((b_{ik})(s)), \ b_{ik}(s) = b_{ki}(s),$$

where $b_{ik}(s)$ *are quadratic forms in* $s = (s_0, s_1, s_2)$. *Moreover, given a sufficiently general sextic* $\Sigma(s)$, *one may reconstruct uniquely the Fano 3-fold as a fibering by conics with the curve* $\Delta_\pi : \Sigma(s) = 0$ *as the base locus of its singular fibers. Further, one may reconstruct the K-3 surface* F' *as a double cover of* $\mathbb{P}^2$ *having the given non-singular sextic branch curve* C *in* $\mathbb{P}^2$.

More generally, we should replace F' by its minimal desingularization, which will then also be a K-3 surface. Thus, one obtains a family of K-3 surfaces, generically as double covers of $\mathbb{P}^2$, and realized as minimal desingularizations of the intersections $Q' \cap S_2$. The generic member F' of this family has $m(F') = 18$ moduli, as we now explain.

A general hyperplane section of the Fano 3-fold is a K-3 surface F' in $\mathbb{P}^7$, realized both as a sextic double cover of $\mathbb{P}^2$ and as a K-3 surface of genus 7 in $\mathbb{P}^7$ whose general hyperplane section in $\mathbb{P}^7$ is a canonically imbedded curve of genus 7 in $\mathbb{P}^6$. Each type of K-3 surface by itself has 19 moduli, but if we consider the family of K-3 surfaces obtained as hyperplane sections of general quadric sections of $\mathbb{P}^2 \times \mathbb{P}^2$ in $\mathbb{P}^8$, then, as S. Tregub observed to me at the Yaroslavl' conference in 1992, and explained in greater detail in a later written communication, pic(F^H) generically has rank 2 and the number of moduli $m(F^H)$ of this kind of K-3 surface F^H is equal to 18.

More importantly, Tregub has explained in the longer written communication [T] how to single out a subfamily of such F^H having 9 moduli. This is important because we should like, if possible, to link this set-up with the second symmetric Hermitian space $\mathbb{H}_3$ and Hermitian modular functions, for the reasons suggested earlier.

The 9-dimensional family described to me by Tregub is the family of K-3 surfaces in our family which may be described as follows. Let $\mathcal{F}$ be the 18-dimensional family of K-3 surfaces F^H described above, let $\mathcal{E}$ be the family of all K-3 surfaces having a fixed-point-free involution, which are therefore the 2-fold unramified covers of Enriques surfaces E. Put $\mathcal{F}_\mathcal{E} = \mathcal{F} \cap \mathcal{E}$. Each of the families $\mathcal{F}$, $\mathcal{E}$, and $\mathcal{F}_\mathcal{E}$ can be characterized by a property of its Hodge structure. Then using this, Tregub shows that $\mathcal{F}_\mathcal{E}$ has a component $\mathcal{F}_{\mathcal{E},0} = M$, say, such that $\dim M = 9$ and that the Enriques surfaces E corresponding to K-3 surfaces from the family M are exactly those which contain a smooth rational curve (cf. [N]). In fact, the family M is generically the family of Reye congruences described by Cossec [Co]. Thus, if F^H is the double cover of an Enriques surface E containing a non-singular rational curve, then $m(F^H) = m(E) = 9$. Such an E is a 2-fold branched covering of a 4-nodal (Cayley) cubic hypersurface $\mathcal{C}$ in $\mathbb{P}^3$ and the branch locus of $\pi : E \longrightarrow \mathcal{C}$ is $C \cup \mathrm{sing}(\mathcal{C})$, where, as I. Dolgachev has informed me, C is one of a determined finite number of smooth curves of genus 4. To be more precise, according to him, there is, in another way, a natural map of degree 24 from the moduli space of nodal Enriques surfaces to M_4, the moduli space of smooth curves of genus 4. He has outlined a proof of this in a longer written communication [D].

Now, first of all, the 9-dimensional moduli space of such nodal Enriques surfaces is a Zariski-open set on an arithmetic quotient of a symmetric tube domain of type IV [N]. Also, a generic smooth curve C of genus 4 is a space sextic realized as $\mathcal{Q} \cap \mathcal{C}$, where the quadric hypersurface $\mathcal{Q}$ is determined up to an element of the finite group of projective automorphisms of $\mathcal{C}$ and the nodal Enriques surface can be recovered from the genus 4 curve $C \subset \mathcal{C}$.

We want to connect naturally the Zariski-open subset of an arithmetic quotient of a domain of type IV and dimension 9 with (a Zariski-open subset of) an arithmetic quotient of the irreducible type I domain $\mathbb{H}_3$ of dimension 9. If Γ_3 is the usual Hermitian modular group, then $\Gamma_3 \backslash \mathbb{H}_3$ parametrizes normally polarized Abelian varieties (A, i) of dimension 6 with complex multiplication by i of type $(i, i, i, -i, -i, -i)$ in the tangent space $T_e(A)$ to A at e. Thus we have the relations or finite correspondences:

$$F^H \longrightarrow E \longrightarrow C \subset \mathcal{C}$$

and we want some finite set of (A, i)'s corresponding to the curve C of genus 4. This brings us to the subject of Janus-type varieties.

(Janus: Ancient Roman god of gates and doorways, depicted with two faces looking in opposite directions.) There are examples of pairs of quite different symmetric domains, say D_1 resp. D_2 on which arithmetic groups Γ_1 resp. Γ_2 operate, such that if $(D/\Gamma)^\#$ is a suitable smooth toroidal compactification of D/Γ, then there are normal crossings divisors Δ_1 and Δ_2 on the respective smooth compactifications $(D/\Gamma_1)^\#$ and $(D/\Gamma_2)^\#$ such that the latter two toroidal compactifications are both isomorphic to a compact projective variety V, and such that

$$V - \Delta_1 = D_1/\Gamma_1, V - \Delta_2 = D_2/\Gamma_2. \tag{1}$$

Such examples are given by B. Hunt in [H] and by Hunt and S. Weintraub in [H-W]. Moreover, there is a natural geometric interpretation of both terms in (1) and the geometric relation is *not random.*

Supposing the genus $g(C) = 4$, let C^{**} be a 4-fold unramified cyclic covering of C, and C^* be the intermediate covering of C of degree 2. There are only finitely many C^{**} for the given C. We have $g(C^{**}) = 13$ and $g(C^*) = 7$. The Jacobian $J(C^*)$ is naturally a subvariety of $J(C^{**})$, and B. van Geemen has shown that

$$A(C^{**}) = J(C^{**})/J(C^*)$$

is a normally polarized Abelian variety of dimension 6 and that the period-four automorphism of C^{**} over C induces complex multiplication of $A(C^{**})$ of type $(i, i, i, -i, -i, -i)$, thus determining an orbit of Γ_3 in $\mathbb{H}_3$. At this point, the first question was whether the correspondence

$$C^{**} \longrightarrow A(C^{**})$$

is generically finite. Now, O. Debarre [De] has given a positive answer to this question, and we have

$$F^H \longrightarrow E \xrightarrow{[24:1]} C \xrightarrow{[1:128]} C^{**} \xrightarrow{\pi} A(C^{**}).$$

One obvious question is to describe the boundary points of the finiteness domain/range of π, as well as the fibers of positive dimension. But from the point of view of our original motivation, we should like to know what kind of analogous phenomena take place in the cases $\mathbb{Q}_3$ and $\mathcal{T}_e$? Might we reasonably, for example, take a generic section of the respective Severi variety in its ambient space $\mathbb{P}^N$ by a smooth quadric variety of suitable co-dimension in $\mathbb{P}^N$? If so, then can we find some objects whose moduli are connected with $\mathcal{T}_e$ contained in the complex Cayley projective plane?

Remark. We describe briefly here how to see that the family of nodal Enriques surfaces or the family of Reye congruences has 9 moduli. Each nodal Enriques surface has as its K-3 covering the minimal desingularization of the double cover of $\mathbb{P}^2$ with a sextic branch curve consisting of two transversal cubic plane curves with a common totally tangent conic (which meets each of the cubics tangentially at three points) [Co]. Since the choice of conic is arbitrary, we can fix it once for all, call it K, and consider the family of all plane cubics to which K is totally tangent. By elementary elimination, one can see that the family of such cubics for the fixed K is six-dimensional. Since one may apply any element of the orthogonal group of K to the plane coordinates, and hence to the two cubics, the number of moduli for the transversal pairs of cubics totally tangent to K is $6+6-3=9$ (3 being the dimension of the complex orthogonal group of K).

Suggested problems

In the case of the orbit space $\mathcal{V}_3 = \Gamma_3\backslash\mathcal{H}_3$ of the Siegel upper half-space $\mathcal{H}_3$ by the Siegel modular group Γ_3, every point $z \in \mathcal{J}_3 = \mathcal{H}_3$ corresponds to a marked curve C of genus 3 which may be reducible, and $\mathcal{V}_3$ is the moduli space of (possibly reducible) curves of genus 3, while a Zariski-open subset $\mathcal{J}_3{}^{\#}$ of $\mathcal{J}_3$ corresponds to *irreducible* curves C of genus 3.

In the same way, the moduli space of nodal Enriques surfaces may be viewed as a Zariski-open set $\mathcal{E}^*$ on an arithmetic quotient $\mathcal{E} \subseteq \Gamma\backslash\mathcal{D}_3$, where $\mathcal{D}_3$ is a type IV domain of dimension 9. Then one should try to characterize the surfaces attached to the points of $\mathcal{E} - \mathcal{E}^*$ (perhaps in a manner analogous to the assignment of reducible curves of genus 3 to the points of $\mathcal{V}_3 - \Gamma_3\backslash\mathcal{J}_3{}^{\#}$).

Moreover, one would like to know more about the correspondence between the quotient of $\mathbb{H}_3$ by the Hermitian modular group and $\Gamma\backslash\mathcal{D}_3$, which Debarre has shown is generically finite: Which 6-dimensional Abelian varieties with multiplication by i correspond to which nodal Enriques surfaces?

One may try to imagine analogous problems in the two higher dimensional cases of $\mathbb{Q}_3$ and $\mathcal{T}_e$, but this is still totally unexplored territory.

References

[Ba1] Baily, W. L., Jr., An exceptional arithmetic group and its Eisenstein series, Ann. Math. 91 (1970), 512–549.

[Be] Beauville, A., Variétés de Prym et Jacobiennes Intermédiaires, Ann. École Norm. Sup. 10 (1977).

[Co] Cossec, F. R., Reye Congruences, Trans. Amer. Math. Soc. 280 (1983), 737–751.

[De] Debarre, O., Surjectivity of a Prym Map, printed communication, Nov. 1993.

[Di] Dixon, A. C., Note on the Reduction of a Ternary Quantic to a Symmetrical Determinant, Proc. Cambridge Philos. Soc. 11 (1902), 350–351.

[D-S] Donagi, R. and Smith, R. C., The Structure of the Prym Map, Acta Math. 146, 25–102.

[D] Dolgachev, I., printed communication dated July 18, 1994

[H] Hunt, B., A Siegel Modular 3-fold that is a Picard Modular 3-fold, preprint, Göttingen 1988 (or Comp. Math. 76 (1990), 203–242).

[H-W] Hunt, B., and Weintraub, S. H., Janus-like Algebraic Varieties, Preprint No. 194, Universität Kaiserslautern 1991.

[Is] Iskovskikh, V. A., Lectures on Three-Dimensional Algebraic Manifolds, Fano Manifolds (in Russian), Matematika, Moscow Univ. 1988.

[Ki] Kim, H., On a Modular Form on the Exceptional Domain of E_7, Thesis, University of Chicago 1992.

[LaZ] Lazarsfeld, R., and Van de Ven, A., Topics in the Geometry of Projective Space, Recent Work of F. L. Zak, with an addendum by F. L. Zak, Birkhäuser, 1984.

[N] Namikawa, Y., Periods of Enriques Surfaces, Math. Ann. 270 (1985), 201–222.

[Tr] Tregub, S., written communication dated Sept. 1992.

[Tyu1] Tyurin, A. N. On Intersections of Quadrics, Russian Math. Surveys 30:6 (1975), 51-105; Uspekhi Mat. Nauk. 30:6 (1975), 51–99.

[Tyu2] Tyurin A. N., The Geometry of the Poincaré Theta-Divisor of a Prym Variety, Izvestia Mat. 39 (5) (1979) (in Russian).

[V] Verra, A., The Prym Map has Degree Two on Plane Sextics, 1992.

[VG] van Geemen, B., Theta functions and cycles on some abelian fourfolds, to appear in Math. Z.

Methods of group analysis in dynamics of non-Newtonian fluid*

V. A. Chugunov, L. D. Eskin, and S. L. Tonkonog

1. Introduction

In this paper, the main results of the study of evolutionary equations describing Glen's flow law fluid sheet dynamics by group analysis are discussed.

Equations of a similar form depict a dynamics of density of turbulent gas being filtrated through a porous medium [1, 2]. Such a system is also faced in the problem on hydro-break through of a ground stratum. From the mathematical point of view, the examined equations represent a wide class of non-linear parabolic equations with singularities of the function and its gradient. All this arouses the mathematical interest in the problem.

2. Results and discussion

2.1. Governing equation

In this section we consider the following equation[1]

$$u_t = (u^2\sigma^n)_x, \quad \sigma = uu_x, \tag{1}$$

describing surface dynamics of a Glen's flow law fluid film moving without slip over a horizontal bed and having no mass origins on its surface. In equation (1) u is the film thickness, t is the time, x is the longitudinal coordinate. In [3, 4] it was shown that equation (1) admits a four-parameter Lie group with the infinitesimal

---

*The authors acknowledge the financial support of the Russian fund of fundamental researches (grant 94-01-01191a).

[1]All the model equations for non-Newtonian fluid dynamics are written for simplicity for an odd number of Glen's flow law exponent. For an arbitrary number n, the equations slightly change [3]. But all the results of this paper remain valid.

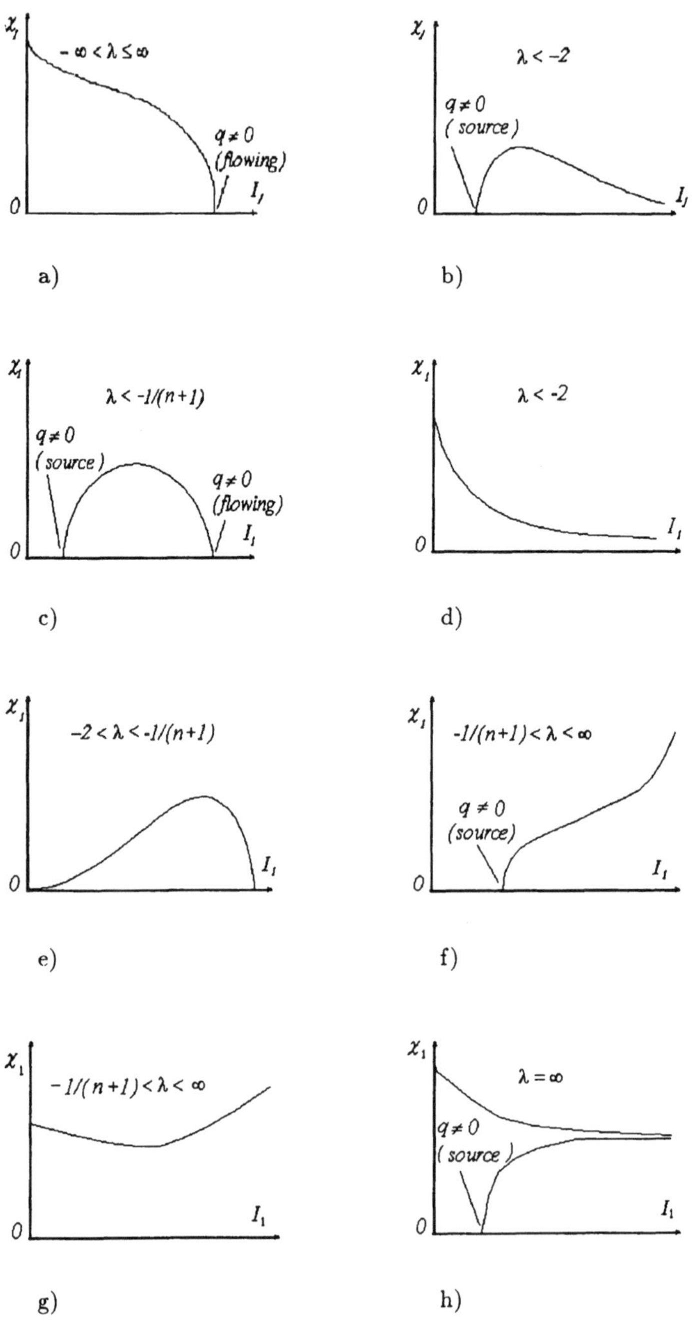

Fig. 1: Classification of the automodel solutions

operators

$$\begin{aligned} X_1^0 &= \partial_t, \\ X_2^0 &= \partial_x, \\ X_3^0 &= t\partial_t + 2x\partial_x + u\partial_u, \\ X_4^0 &= (n+1)t\partial_t + x\partial_x - \sigma\partial_\sigma. \end{aligned}$$

Automodel solutions of equation (1) are invariant with respect to the group $X^0(\lambda) = X_3^0 + \lambda X_4^0$, and its functional-independent invariants are

$$I_1 = xt^{-b_1}, \quad I_2 = ut^{-b_2}, \quad I_3 = \sigma t^{\lambda b_2},$$
$$b_1 = (2+\lambda)b_2, \quad b_2 = (1+\lambda(n+1))^{-1}, \tag{2}$$
$$\lambda \neq -(n+1)^{-1}.$$

Then the required automodel solutions take the following form:

$$u = t^{b_2}\chi_1(I_1), \quad \sigma = t^{-\lambda b_2}\chi_2(I_1),$$

where the functions χ_1, χ_2 satisfy the equation

$$(\chi_1^2\chi_2^n)' + b_1(\lambda)I_1\chi_1' - b_2(\lambda)\chi_1 = 0, \chi_2 = \chi_1\chi_2'. \tag{3}$$

When $\lambda = -3$, equation (3) has the solution

$$u = c_n t^{b_2}(\xi_0^r - I_1^r)^p, \quad \xi_0 - \text{const}, \tag{4}$$
$$c_n = (-b_2)^{p/n}(pr)^{-p}, \quad r = (n+1)/n, \quad p = n/(2n+1).$$

For solution (4) we have $q_{x=0} = 0$ and $q_{x=x_f} = 0$, where q is the fluid massflux, $x_f = \xi_0 t^{b_1}$ the terminus coordinates such that $u(t, x_f) = 0$. The solution (4) describes a free (with zero mass balance) spreading of the film. The other interesting solution of equation (1) is the automodel solution, which is invariant with respect to the operator X_4^0 ($\lambda = \infty$). This solution has the form $u = \chi_1(xt^{-b_1})$, satisfies equation (3) when $b_1 = (n+1)^{-1}$, $b_2 = 0$ and is completely determined by imposing $u_0 = \chi_1(0)$ and equating incharge at the terminus to zero. This solution describes stretching out of the fluid film flowing down from a lap situated in the plane $x = 0$ on the height u_0. When $\lambda = -2$, equation (3) can be integrated in quadrants. The solution of equation (1), which is invariant with respect to the operator $X^0(-1/(n+1))$, can be found in closed form.

Equation (3) can be transformed to a set of equations in the plane for new unknown functions by substitutions

$$\varphi(\eta) = I_1^{-\frac{n+1}{2n+1}}\chi_1, \quad \psi(\eta) = \varphi'(\eta), \quad \eta = \ln I_1.$$

With the help of the qualitative theory of differential equations, it is possible to completely classify both all the single-parameter families of solutions to this system and separatrises separating it. Whence, overall classification of all solutions to equation (3) and, therefore, of all automodel solutions to equation (1) can be determined [5]. In Fig.1, all the two-parameter families (which are nine) of solutions to equation (3) descending from the single-parameter families of solutions to the set of equations are depicted. There are yet six single-parameter families of solutions to equation (3), which come from the separatrises. Both solution (4) and the solution describing film spreading from a lap belong to the last.

Asymptotics of the curve $\chi_1 = \chi_1(I_1)$ near the margin $\xi_0 \neq 0$ when $q \neq 0$ is determined by relations:

$$\chi_1 \sim c(\xi_0 - I_1)^{\frac{n}{2(n+1)}}, \quad I_1 \to \xi_0 - 0, \quad q < 0$$

(discharge on the terminus)

$$\chi_1 \sim c(I_1 - \xi_0)^{\frac{n}{2(n+1)}}, \quad I_1 \to \xi_0 + 0, \quad q > 0$$

(incharge on the terminus).

The constant $c > 0$ is determined from the choice of the curve. In the case of zero discharge on the terminus, the exponent $\frac{n}{2(n+1)}$ in asymptotics of the function χ_1 is replaced by $\frac{n}{(2n+1)}$.

For infinite solutions, its asymptotics, when $I_1 \to \infty$, has the form $\chi_1 \sim cI_1^{\frac{1}{2+\lambda}}$.

2.2. Disturbed equation

Let us consider the disturbed equation

$$u_t = (u^2\sigma^n)_x + \epsilon\Phi(t, x, u, \sigma), \quad \sigma = uu_x, \tag{5}$$

describing dynamics of the film surface with the power of surface origin $\epsilon\Phi$. Below, overall classification of equations (5) admitting an approximate group of symmetry in the sense of work [6] is proposed.

Let us denote

$$\varsigma_1 = \frac{(n+1)\theta_1 - \varphi'}{2n+1}, \quad \varsigma_2 = \frac{\theta_1 - 2\varphi'}{2n+1}, \quad \zeta_3 = \int t^{-b_1(\lambda)}\theta_2' dt,$$

where φ, θ_1, θ_2 are arbitrary functions of the independent variable t.

Let us introduce

$$X^1 = \varphi\partial_t + (\theta_1 x + \theta_2)\partial_x + \varsigma_1 u\partial_u + \varsigma_2\sigma\partial_\sigma.$$

It is shown in [7], that there exist only four classes of equation (5) admitting to order $O(\epsilon^2)$ an approximate group of symmetry, namely

a) the equation

$$u_t = (u^2\sigma^n)_x + \epsilon[\varsigma_1 u - (\theta_1 x + \theta_2)\sigma u^{-1} + f(x, u, \sigma)], \tag{6}$$

$$\sigma = uu_x,$$

is approximate-invariant with respect to the group with the operator $X_1 = X_1^0 + \epsilon X^1$;

b) the equation

$$u_t = (u^2\sigma^n)_x + \epsilon[\varsigma_1' xu - (0.5\theta_1' x^2 + \theta_2' x)\sigma u^{-1} + f(t,u,\sigma)], \tag{7}$$

$$\sigma = uu_x,$$

is approximate-invariant with respect to the approximate group with the operator $X_2 = X_2^0 + \epsilon X^1$;

c) the equation

$$u_t = (u^2\sigma^n)_x + \epsilon[b_2 t^{-1}(\varsigma_1 u - (\theta_1 x + t^{b_1}\varsigma_3)\sigma u^{-1}) + t^{-\lambda(n+1)b_2} f(I_1, I_2, I_3)], \tag{8}$$

$$\sigma = uu_x,$$

is approximate-invariant with respect to the approximate group with the operator $X(\lambda) = X^0(\lambda) + \epsilon X^1$;

d) the equation

$$u_t = (u^2\sigma^n)_x + \epsilon\left[\varsigma_1 u - \left(\theta_1 x + \theta_2 + \lambda\int \theta_1 dt)\sigma u^{-1} + f(x + \lambda t, u, \sigma\right)\right], \tag{9}$$

$$\sigma = uu_x$$

is approximate-invariant with respect to the approximate group with the operator $X = \partial_t - \lambda\partial_x + \epsilon X^1$.

In equations (6)–(9), f is an arbitrary function of its arguments, the functions I_1, I_2, I_3 are determined by relation (2).

For any of the equations (6)–(9) the approximate-invariant solutions are found to order $O(\epsilon^2)$ [7]. As an example, the approximate-invariant solution of equation (8), when $\theta_2 = 0$ can be shown. It has the form ([7])

$$u = \chi_1(1 + \epsilon\nu_1)\exp\left[\int (1 + \epsilon\varsigma_1)\alpha^{-1} dt\right],$$

$$\sigma = \chi_2(1 + \epsilon\nu_2)\exp\left[\int (-\lambda + \epsilon\varsigma_2)\alpha^{-1} dt\right],$$

where $\alpha = b_2^{-1} t + \epsilon\varphi$.

The functions χ_1, χ_2 satisfy equation (3), in which the argument I_1 is replaced by $J = x\exp[-\int(2 + \lambda + \epsilon\theta_1)\alpha^{-1}dt]$. The invariant of the approximate group of symmetry for equation (8) and the functions $\nu_1(J)$, $\nu_2(J)$ satisfy the set of linear equations

$$b_2^{-1}[\chi_1\nu_1 - (2+\lambda)J(\chi_1\nu_1)'] - [\chi_1^2\chi_2^n(2\nu_1 + n\nu_2)]' = f(J, \chi_1, \chi_2),$$

$$2\nu_1 + \chi_1^2\chi_2^{-1}\nu_1' = \nu_2.$$

Analogous results are valid for the approximate-invariant solutions of the equations (6), (7), (9). But, in these cases functions χ_1, χ_2, ν_1, ν_2 can be found in closed forms.

2.3. Exact symmetries of disturbed equations

Determination of the conditions providing an accuracy of the approximate symmetries for equtions (6)–(9) shown in the third section represents the next problem. For equation (7) and the operator X_2 we have the following result (S. L. Tonkonog, L. D. Eskin):

Theorem. *In order to equation* (7) *be exact invariant with respect to the point group having the operator* X_2, *it is necessary and sufficient that the functions* φ, θ_1, θ_2 *satisfy the set of equations*

$$\begin{aligned}(\varphi\varsigma_1')' + \theta_1\varsigma_1' &= 0,\\ (\varphi\theta_1')' + \theta_1\theta_1' &= 0,\\ (\varphi\theta_2')' + \theta_2\theta_1' &= 0,\end{aligned} \tag{10}$$

and the function $f(t,u,\sigma)$ *satisfies the equation*

$$\varphi f_t + \varsigma_1 u f_u + \varsigma_2 \sigma f_\sigma - (n+1)\varsigma_2 f = \theta_2(\theta_2'\sigma u^{-1} - \varsigma_1' u).$$

Analogous results are valid for the equations (6), (8), (9), and besides, in the case of equation (8) the set of equations (10) must be replaced by a set of integral-differential equations for the functions φ, θ_1, θ_2. This theorem and its analogs for the equations (6), (8), (9) permit to build a lot of equations (5) admitting, unlike equation (1), exact groups of symmetry different from stretching.

Let us restrict ourselves to two examples:

1). Considering the solution of equation (10)

$$\varphi = t^{\frac{2n+2}{2n+3}}, \quad \theta_1 = \frac{2}{2n+3} t^{-\frac{1}{2n+3}}, \quad \theta_2 = 0,$$

with the help of the theorem, we find that the following equation

$$u_t = (u^2\sigma^n)_x + \epsilon\left[(2n+3)^{-2}x^2 t^{-\frac{2n+4}{2n+3}}\sigma u^{-1} + t^{-\frac{2n+2}{2n+3}}\psi(u, t^{\frac{2}{2n+3}})\right], \tag{11}$$

$\sigma = u u_x$ (ϵ is arbitrary constant, ψ is arbitrary function of two variables) is invariant with respect to the group with the operator

$$X_2 = \epsilon t^{\frac{2n+2}{2n+3}}\partial_t + \partial_x + \frac{2\epsilon}{2n+3} t^{-\frac{1}{2n+3}}(x\partial_x - \sigma\partial_\sigma). \tag{12}$$

Invariant with respect to the group with the operator (12), the solution of equation (11) takes the form

$$u = \chi_1(J_1), \quad \sigma = t^{-\frac{2}{2n+3}}\chi_2(J_1),$$

where

$$J_1 = x t^{-\frac{2}{2n+3}} + \epsilon^{-1}(2n+2)t^{-\frac{1}{2n+3}}$$

is the invariant of group (12), and the functions χ_1, χ_2 satisfy equation (3).

2). The equation

$$u_t = (u^2\sigma^n)_x + \epsilon \left[b_2(\lambda) t^{k-2} \left(\frac{n+1}{2n+1} u - x\sigma u^{-1} \right) + t^{-\lambda(n+1)b_2} \psi \left(I_2 I_1^{-\frac{n+1}{2n+1}}, I_3 I_1^{-\frac{1}{2n+1}} \right) \right], \tag{13}$$

$$\sigma = u u_x$$

(ψ is an arbitrary function) can be found from equation (8), when $\varphi = 0$, $\theta_1 = t^{k-1}$, $\theta_2 = 0$. It is the exact invariant with respect to the group with the operator

$$X(\lambda) = t b_2^{-1} \partial_t + (2 + \lambda + \epsilon t^{k-1}) x \partial_x + \left(1 + \epsilon \frac{n+1}{2n+1} t^{k-1} \right) u \partial_u + \left(-\lambda + \frac{\epsilon}{2n+1} t^{k-1} \right) \sigma \partial_\sigma. \tag{14}$$

Let us denote

$$J_1 = x t^{-b_1} \exp[-(2n+1)\mu(t)]$$

the invariant of the group (14), where

$$\mu(t) = \epsilon b_2 [(k-1)(2n+1)]^{-1} t^{k-1}.$$

It is can be shown that the invariant solution of equation (13) has the form

$$u = \chi_1(J_1) t^{b_2} \exp[(n+1)\mu], \quad \sigma = \chi_2(J_1) t^{-\lambda b_2} \exp(\mu), \tag{15}$$

where χ_1 and χ_2 satisfy the following equation:

$$(\chi_1^2 \chi_2^n)' + b_1 J_1 \chi_1' - b_2 \chi_1 + J_1^{\frac{n+1}{2n+1}} \psi \left(\chi_1 J_1^{-\frac{n+1}{2n+1}}, \chi_2 J_1^{-\frac{1}{2n+1}} \right) = 0, \tag{16}$$

$$\chi_2 = \chi_1 \chi_1'.$$

When $\psi = 0$, equation (16) coincides with equation (3), where the argument I_1 is replaced by J_1. However, physical pictures of the film flows described by the invariant solutions (15) of the disturbed equation (13) essentially differ from those depicted by the automodel solutions of the undisturbed equation (1). Indeed, when $\lambda = -3$ and $\psi = 0$, the invariant (automodel) solution of equation (1) is determined by relation (4), which describes a free spreading of the film, and besides $x \to \infty$ as $t \to \infty$. The invariant solution of equation (13), when $\lambda = -3$, $\psi = 0$ and $k \neq 1$, due to (15) and equation (16), takes the form

$$u = c_n t^{-\frac{1}{3n+1}} \exp \left[-\frac{\epsilon(n+1)t^{k-1}}{(3n+2)(2n+1)(k-1)} \right] (\xi_0^{\frac{n+1}{n}} - J_1^{\frac{n+1}{n}})^{\frac{n}{2n+1}}, \tag{17}$$

$$J_1 = x t^{-\frac{1}{3n+2}} \exp \left[\epsilon \frac{t^{k-1}}{(3n+1)(k-1)} \right],$$

hence

$$\max_x u(x,t) = u(0,t) = c_n \xi_0^{\frac{n+1}{2n+1}} t^{-\frac{1}{3n+1}} \exp \left[-\frac{\epsilon(n+1)t^{k-1}}{(3n+2)(2n+1)(k-1)} \right],$$

$$x_f(t) = \xi_0 t^{\frac{1}{3n+1}} \exp\left[-\epsilon \frac{t^{k-1}}{(3n+1)(k-1)}\right].$$

a) In the case of $k > 1$, when $0 < t \leq \epsilon^{-\frac{1}{k-1}}$ $(\epsilon > 0)$, x_f increases from 0 to

$$x_f^{\max} = \xi_0 \epsilon^{-\frac{1}{(3n+2)(k-1)}} \exp\left[-\frac{1}{(3n+2)(k-1)}\right]$$

and the film stretches, as in the case of solution (4) of equation (1). Then, as t increases, $x_f(t)$ decreases and $x_f(t) \to 0$, $u(0,t) \to 0$ as $t \to \infty$, that is, the film degrades and solution (17) simulates, for example, advance and retreat of a glacier. This is explained by the fact that, at first, the film dynamics is caused by the rheological term (first term in the right-hand side of equation (13)), predominating influence of which, as in the case of equation (1), leads to the film spreading. Then, as t increases, when $k > 1$, the process is caused by the second term (mass balance) in the right-hand side of equation (13), which is negative for solution (17) and plays a part of surface mass discharge increasing when $k > 2$ and slowly descreasing when $1 < k < 2$. This arouses degradation of the film.

b) When $k < 1$, then $x_f(t) \to \infty$, $u(0,t) \to 0$ as $t \to \infty$ and the film stretches. This is caused by fast decrease of the second terms value in the right-hand side of (13), which models a surface mass discharge as t increases. But the impact of this outflux decreases and flow essence starts to be in much more extent determined by the rheological term. The solution (17) approaches to the automodel solution (4) of equation (1) as t increases.

c). When $k = 1$, then, instead of formula (15) for the invariant solution of equation (13), we derive

$$u = t^{[1+\frac{\epsilon(n+1)}{2n+1}]b_2}\chi_1(J_1), \quad \sigma = t^{(-\lambda+\frac{\epsilon}{2n+1})b_2}\chi_2(J_1), \quad J_1 = xt^{-(2+\lambda+\epsilon)b_2}. \tag{18}$$

When $\lambda = -3$ and $\psi \equiv 0$, then for invariant solution (18) we have

$$x_f = \xi_0 t^{\frac{1-\epsilon}{3n+2}}, u(0,t) = c_n \xi_0^{\frac{n+1}{2n+1}} t^{-\beta},$$

where $\beta = (1 + \epsilon\frac{n+1}{2n+1})(3n+2)^{-1}$.

Hence we can conclude that, when $\epsilon > 1$, the film degrades, when $\epsilon = 1$, the margin is stationary, and, when $0 < \epsilon < 1$, the fluid spreads.

2.4. Equation with regard to slipping effect

An equation of the following form,

$$u_t = (q^\epsilon)_x, \tag{19}$$

where fluid flux $q^\epsilon = u^{n+2}u_x^n + \epsilon u^{m+1}u_x^m$, $m < n$, $\epsilon << 1$ emerges in the modelling of the surface dynamics of a film moving with slip over a horizontal bed [8].

The great interest is aroused by the problem of free spreading with sliding, in which we have following boundary and initial conditions:

$$x = 0 : q^\epsilon = 0; x = x_f(t) : q^\epsilon = 0; t = 0 : u = M\delta(x), \tag{20}$$

where M is the film mass per unit width, $x_f(t)$ the unknown terminus. In free spreading, because of absence of the surface outflux, its overall mass remains unchanged, whence

$$\int_0^{x_f} u dx = 2^{-1}M. \tag{21}$$

Equation (19) as it has been shown in [3] admits a group with the operator

$$Y = (m - 2n - 1)t\partial_t + (2m - 2n - 1)x\partial_x + (m - n)u\partial_u.$$

However, solution invariant with respect to this group does not satisfy the condition of constant mass and, hence, is not a solution of free spreading with slip.

In [9] we proposed an algorithm for solving the problem (19), (20) to order $O(\epsilon^2)$, based on the fact that equation (19) admits a group of symmetry, transforming both the variables t, x, u, and the parameter ϵ, namely the group with the following operator

$$X(\epsilon) = -(3n + 2)t\partial_t - x\partial_x + u\partial_u + [3(n - m) + 1]\epsilon\partial_\epsilon$$

(compare with the operator $X^0(\lambda)$, $\lambda = -3$ in section 2.1).

An X-invariant solution of equation (19) has the form

$$u = t^{-\frac{1}{3n+2}} \tilde{\chi}(\xi, \eta),$$

where $\xi = I_1 = xt^{-\frac{1}{3n+2}}$, $\eta = \epsilon t^{\frac{1+3(n-m)}{3n+2}}$ are the invariants of the group $X(\epsilon)$.

The formula for the margin movement in new variables ξ, η is written as $\xi_0 = g(\eta)$ $(\tilde{\chi}(\xi, \eta) = 0)$. The functions $\tilde{\chi}$, g are unknown and must be determined. It is useful to introduce a new variable $z = \xi/g(\eta)$ instead of ξ, then $z_f = 1$ (terminus is stationary). For the function $\chi(z, \eta) = \tilde{\chi}(\xi, \eta)$ we obtain, due to (19), the relation

$$\chi + z\chi_z - [1 + 3(n - m)]\eta(\chi_\eta - z\chi_z g g_\eta^{-1}) = -(3n + 2)(q^\epsilon)_z, \tag{22}$$

where $q^\epsilon = q^\epsilon(\chi, \chi_z, g, \eta)$.

The solution of equation (19) of order $O(\epsilon^2)$ with regard to boundary and initial conditions (20) can be chosen in the form

$$\chi = v(z) + \eta w(z) + O(\eta^2), \quad g = a + b\eta + O(\eta^2). \tag{23}$$

The functions v, w and the constants a, b are unknown and should be determined. Substituting (23) into (22), (20) and (21) and equating factors at η^0 and η^1 we derive a non-linear equation for $v(z)$ and a linear differential equation of second order for $w(z)$ with coefficients expressed by v, v_z, v_{zz}, a, b (b enters the equation linearly. Details can be found in [9]). To the second order of η, $v(z)$ can be found in completed form of type (4). The constant a can be determined from the mass constancy (21). Coefficients of the linear differential equation for $w(z)$ have the algebraic singularity (the point of branching) when $z = 0$ and $z = 1$. Asymptotic analysis of its solution in the vicinity of $z = 1$ shows that there exist its single solution represented by the series

$$w = (1-z)^{\gamma} \sum_k c_k (1-z)^k, \quad \gamma = \frac{n-m}{2n+1} > 0,$$

including the parameter b and satisfying boundary condition when $z = 1$, arising from absence of marginal flux. Parameter b can be found from the boundary condition when $z = 0$ (absence of flux in divide $x = 0$). In doing so, it is proved to be handy to use the linearity of the equation for $w(z)$ with respect to b. Performed numerical calculations show that $b > 0$ and that Kamb's model really describes the effect of the film slip over the bed.

References

[1] Barenblatt, G. I., Some Non-Steady Motions of the Liquid and gas in the Porus Medium, Prikl. Mat. Mekh. 16 (1) (1952).

[2] Barenblatt, G. I., Automodel Motions of the Compresssible Fluid in the Porus Medium, Prikl. Mat. Mekh. 16 (6) (1952).

[3] Salamatin, A. N., Chugunov, V. A., and Mazo, A. B., Numerical Investigation and Invariant Solutions of the Problem of Subisothermal Glacier Dynamics in the Two-Dimensional Approximation, Problemy Mekhaniki Prirodnykh Processov, Grigorian S. S. and Krass M. S., eds., Moscow Univ., Moscow 8, 1983.

[4] Chugunov, V. A., Group Properties of the Equation Describing glacier flow, Izv. Vuzov. Matematika 10 (1982), 84.

[5] Tonkonog, S. L., Automodel Solutions of the Equation Describing Glacier Flow, Issled. Prikl. Mat. 21 (1995).

[6] Baikov, V. A., Gazizov, R. K., and Ibragimov, N. H., Approximate Symmetries, Mat. Sb. 136 (3) (1988), 435.

[7] Tonkonog, S. L., Approximate Group Analysis of the Equations of Subisothermal Glacier Dynamics, Izv. Vuzov. Matematika 33 (4) (1994).

[8] Kamb, B., Sliding Motion of Glaciers: Theory and Observation, Reviews of Geophysics and Space Physics 3 (4) (1970), 673.

[9] Tonkonog, S. L., Chugunov, V. A., and Eskin, L. D., Methods of Group Analysis for the Problem of a Movement of the Thin Sheet of Non-Newtonian Fluid with regard of Slipping Effect, Prikl. Mat. Mekh. 58 (4) (1994), 63.

Discontinuous phenomena and Turing definability*

S. Barry Cooper

Abstract. We discuss the relationship between discontinuity and definability in the Turing degrees, with particular reference to degree invariant solutions to Post's Problem. Proofs of new results concerning definability in lower cones are outlined.

1. Introduction

The work of Gödel [1931], [1934], Turing [1936], Church [1936] and others in the 1930's showed that in mathematics many basic decision problems can be reduced to those for particular noncomputable, computably enumerable (c.e.) sets. However (see for example Myhill [1955]) particular examples tend to lie in $\mathbf{0}'$ (the Turing degree of first-order Peano arithmetic and of the halting problem for Turing machines). So it was natural to ask whether there are other c.e. degrees $> \mathbf{0}$ (the degree of the computable sets).

Post's Problem 1.1 (Post [1944]). Does there exist a property P of (indices) of c.e. sets such that

$$e \in P \Rightarrow \emptyset <_T W_e <_T \emptyset'?$$

(See Soare [1987] or Odifreddi [1989] for basic terminology and notation.)

An affirmative answer to this question was independently provided by Friedberg [1957] and Mučnik [1956]. But the property P was not of the form originally envisaged by Post, $P(e)$ saying "e is the index corresponding to a particular finite-injury priority construction". So the question arises as to how *natural* a property we can expect P to be. That is, given that the Friedberg–Mučnik theorem shows the existence of intermediate c.e. degrees, does there exist a description of such a class of degrees which invests them with more than a pathological significance?

*Preparation of this paper partially supported by a Royal Society travel grant and E.P.S.R.C. Research Grant no. GR/H 02165. We also wish to acknowledge helpful discussions with L. Harrington, R. A. Shore, T. A. Slaman and X. Yi (made possible by visits variously supported by E.P.S.R.C. research grants nos. GR/J28018 and GR/H91213, and a Nuffield Foundation travel grant.)

One may distinguish three current criteria for describing P to be "natural":

(a) P might be of the form "e is the index of a particular naturally occuring c.e. set". That is, P might show some naturally occurring class of sets in mathematics to have members of intermediate c.e. degree.

There are some striking results in this direction. Feferman [1957] showed that every c.e. degree contains a formal system (which, by Shoenfield [1958], can be chosen to be essentially undecidable), and even (Hanf [1965]) contains a *finitely* axiomatisable first-order theory. Every consistently axiomatisable theory has a model of degree $< \mathbf{0}'$ (Shoenfield [1960]); Matijasevič [1970] et al, in solving Hilbert's tenth problem, showed that every c.e. set is diophantine.

(b) The description P may in itself may be particularly simple or natural in form (that is, a P in the spirit of Post's original question).

The first such example is due to Marchenkov [1976], followed by other increasingly convincing solutions to Post's problem in its original form. Most conclusive is the Harrington–Soare [1991] description of such a P in the theory of the lattice of c.e. sets.

Finally, extending criterion (b) to include descriptions of P framed within the language of computability theory itself:

(c) P may be recursion theoretically *robust*.

There are two closely related open questions with claims to provide such a property P.

Question 1.2 (Sacks [1967]). Is there a degree invariant solution to Post's problem? That is, is there an index e such that for any $A, B \subseteq \omega$ we have

$$A <_T W_e^A <_T A'$$

and

$$A \equiv_T B \Rightarrow W_e^A \equiv_T W_e^B?$$

Question 1.3 (Slaman, see Sacks [1985]). Is there an intermediate c.e. $\mathbf{a}$ which is *definable* (in the language of the theory of $\mathfrak{C}$, the structure of the c.e. degrees)?

A positive solution to Sacks' longstanding question would be provided by any *relativisable* construction of a definable intermediate c.e. $\mathbf{a}$. Of course, there are definitions of $\mathbf{a}$ outside the theory of $\mathfrak{C}$ which would suffice for 1.2.

Sacks' question is not just of interest to computability theorists, but is closely connected to two very general conjectures in set theory (see Slaman [ta]).

Let $\mathcal{F}$ be the class of all degree invariant functions from 2^ω to 2^ω, with the ordering $\leq_M$ given by:

$$f \leq_M g \Leftrightarrow (\exists X)(\forall Y \geq_T X)[f(Y) \leq_T g(Y)].$$

(That is, $f \leq_M g$ if $f \leq_T g$ a.e.)

Martin's Conjecture 1.4. Assume AD. Then

(a) If $f \not\geq_M$ id then f is constant a.e.

(b) On those elements of $\mathcal{F}$ which are $\geq_M$ id, $\leq_M$ is a quasiwellordering with successor equal to the Turing jump.

We say that a Borel equivalence relation U on Polish spaces, with countable equivalence classes, is *universal* if and only if for every other such equivalence relation E there is a Borel function f such that for all x and y

$$x \sim_E y \Leftrightarrow f(x) \sim_U f(y).$$

Kechris's Conjecture 1.5. *$\equiv_T$ is universal.*

Slaman and Steel [1988] have shown that arithmetic equivalence is universal. Roughly speaking, Martin's conjecture says that (assuming AD) the only nontrivial, definable, degree invariant operators are the known jump operators and their iterates. It implies the failure Kechris's conjecture and (in particular) a negative answer to Sack's question.

We note that for a given index e, any counterexample A, B to 1.2 may have to be quite complicated.

Theorem 1.6 (Slaman and Steel, unpublished). *There is an index e such that for any $A, B \in \Delta^0_2$ we have*

$$A <_T W_e^A <_T A'$$

and

$$A \equiv_T B \Rightarrow W_e^A \equiv_T W_e^B.$$

Also, Lachlan [1975a] showed that there is no index e as in question 1.2 for which the Turing reductions between W_e^A and W_e^B can be obtained *uniformly* from those for the reductions between A and B.

Any discussion of definability of computably enumerable Turing degrees is incomplete without some mention of the notion of *bi-interpretability*. The Bi-Interpretability Conjecture for $\mathcal{C}$ (due to Harrington and Slaman) posits the existence of a kind of pseudo-isomorphism between ω (the standard model of first-order arithmetic) and $\mathcal{C}$, sufficient to invest $\mathcal{C}$ with some familiar properties of ω (such as rigidity, definability of arithmetically describable sets – in particular, of all individual members of the domain). Roughly speaking, a bi-interpretation between ω and $\mathcal{C}$ consists of:

(1) A coding of ω into $\mathcal{C}$ (involving specifying a collection of c.e. degrees, and relations on those degrees which represent addition and multiplication),

(2) A mapping of each $x \in \mathcal{C}$ to an index e for a representative W_e of $\mathbf{x}$, and

(3) A uniform definition in $\mathcal{C}$ of the relationship between any $\mathbf{x}$ and the code of the index e for the representative W_e of $\mathbf{x}$.

The best result in this direction so far is:

Theorem 1.7 (Harrington and Slaman). Th($\mathcal{C}$) $\equiv_{1-1}$ *the first-order theory of arithmetic (and so* $\in \mathbf{0}^{(\omega)}$*).*

Unfortunately, a proof of bi-interpretability would not necessarily solve question 1.3 so as to give a degree invariant solution to Post's problem. The definition of any intermediate c.e. degree **x** would just describe (degree theoretically) the index e for $W_e \in \mathbf{x}$, and requiring this to give us a relativisable definition of an intermediate c.e. degree effectively returns us to the original question.

2. Continuity in $\mathcal{C}$

A first step in the search for a *natural* definition of a singleton in $\mathcal{C}$ must be to consider questions of the form:

- Is there a definable $P \subset \mathcal{C} - \{\mathbf{0}, \mathbf{0}'\}$ with a *least* or *greatest* member? Or even:
- Is there such a P with *minimal* or *maximal* elements?

In doing this one must take account of known *continuity* properties of $\mathcal{C}$.

Let $F(x,y)$ be a formula in the language $\mathcal{L}$ for $\langle \mathcal{C}, \leq, \cup, \cap, \mathbf{0}, \mathbf{0}' \rangle$. The strongest such continuity notion (see Harrington and Soare [ta]) is:

Definition 2.1. If $\mathbf{a}, \mathbf{b} \in \mathcal{C}$, we say $F(x,y)$ *holds in a neighbourhood of* $\mathbf{a}, \mathbf{b}$ if there exist c.e. degrees $\mathbf{a}_0, \mathbf{a}_1, \mathbf{b}_0, \mathbf{b}_1$ with $\mathbf{a} \in (\mathbf{a}_0, \mathbf{a}_1)$ and $\mathbf{b} \in (\mathbf{b}_0, \mathbf{b}_1)$ such that $F(\mathbf{x}, \mathbf{y})$ holds for all c.e. $\mathbf{x} \in (\mathbf{a}_0, \mathbf{a}_1)$ and $\mathbf{y} \in (\mathbf{b}_0, \mathbf{b}_1)$.

Then:

Theorem 2.2 (Harrington–Soare). *Let* $F(x,y)$ *be any open formula of* $\mathcal{L}$ *with two variables, such that* $F(\mathbf{a}, \mathbf{b})$ *holds for c.e. degrees* $\mathbf{a} \neq \mathbf{b}$*. Then* $F(x,y)$ *holds in a neighbourhood of* $\mathbf{a}, \mathbf{b}$.

Proof. This reduces to considering four basic cases:

1) If $F(\mathbf{x}, \mathbf{y})$ says that $\mathbf{0} < \mathbf{x} < \mathbf{y} < \mathbf{0}'$, then the required intervals $(\mathbf{a}_0, \mathbf{a}_1)$, $(\mathbf{b}_0, \mathbf{b}_1)$ are easily got via the Sacks [1964] Density Theorem.

2) $F(\mathbf{x}, \mathbf{y})$ says that $\mathbf{x}, \mathbf{y}$ form a minimal pair (with $\mathbf{x} \cup \mathbf{y} < \mathbf{0}'$ by the Lachlan [1966] Nondiamond Theorem). Then the result follows using the Harrington–Soare [ta] 'continuity of capping' theorem to get $\mathbf{a}_1, \mathbf{b}_1$ and Sacks Density to get $\mathbf{a}_0, \mathbf{b}_0$.

3) $F(\mathbf{x}, \mathbf{y})$ says that $\mathbf{x}, \mathbf{y}$ form a splitting of $\mathbf{0}'$. This is the dual of case 2, here using the 'continuity of cupping' result of Ambos-Spies, Lachlan and Soare [1993].

4) $F(\mathbf{x}, \mathbf{y})$ says $\mathbf{x} \mid \mathbf{y}$ with $\mathbf{x} \cup \mathbf{y} \neq \mathbf{0}'$ and $\mathbf{x} \cap \mathbf{y} \neq \mathbf{0}$. In this case, one gets $\mathbf{a}_0, \mathbf{b}_0$ by applying the Robinson [1971] Low Splitting Theorem to $\mathbf{a}$ and $\mathbf{b}$ over some low $\mathbf{d} > \mathbf{0}$, incorporating avoidance of the appropriate lower cone below $\mathbf{a}$ or $\mathbf{b}$. And then get $\mathbf{a}_1, \mathbf{b}_1$ by the Robinson [1971] extension of Sacks Density applied

to intervals $(\mathbf{a}, \mathbf{a} \cup \mathbf{b})$ and $(\mathbf{b}, \mathbf{a} \cup \mathbf{b})$, avoiding the respective upper cone above $\mathbf{b}_0$ or $\mathbf{a}_0$. □

Athough this general continuity result cannot be extended to open formulas with more than two variables, or to formulas with quantifiers, further examples of continuity include (1) (Seetapun [ta]) the continuity of Lachlan [1979] nonbounding, and (2) (arising from recent work of Cooper, Lachlan, Seetapun, Slaman and Yi) the continuity of Harrington nonsplitting base.

Weaker notions of continuity (framed in terms of limits) provide further limitations on such definable properties P. One of the longest-standing continuity problems, that concerning the existence of *major subdegrees* (see Harrington and Soare [ta]) for intermediate c.e. degrees, asks about lower continuity of $\varphi(a, x)$, where a is a parameter and

$$\varphi(\mathbf{a}, \mathbf{x}) \Leftrightarrow \forall \mathbf{y}(\mathbf{a} \cup \mathbf{y} = \mathbf{0}' \Leftrightarrow \mathbf{x} \cup \mathbf{y} = \mathbf{0}').$$

Definability relative to other structures looks more promising in that there are known nontrivial discontinuity properties. For instance density fails in both the d.r.e. degrees $\mathbf{D}_2$ (Cooper–Harrington–Lachlan–Lempp–Soare [1991]) and in the enumeration degrees (see Cooper [1990] or Calhoun–Slaman [ta]). See Arslanov, Lempp and Shore [1996] for results relating to definability of c.e. degrees in $\mathbf{D}_2$.

3. Discontinuity and definability

Initial segment results (see Lerman [1983]) provided the technical background to early definability results such as those (over $\mathfrak{D}'$) of Jockusch and Simpson [1976], and:

Theorem 3.1 (Simpson [1977]). *The degree of* Th($\mathfrak{D}$) *(= the first-order theory of $\mathfrak{D}$) = the degree of* Th($\mathcal{N}$) *(the theory of second-order arithmetic).*

These have largely been superseded by the technically simpler ad hoc coding techniques of Nerode and Shore [1980], [1980a], and more recently of Slaman and Woodin [1986] (see also Slaman [1991]), giving rise to an impressive body of definability results for $\mathfrak{D}$ and other degree structures. Examples in addition to Theorem 1.7 above are given by:

Theorem 3.2. (1) (Slaman–Woodin [ta]) $\mathbf{0}''$ *is definable in* $\mathfrak{D}$.

(2) (Shore [1988]) High$_n$, Low$_n$ *is definable in* $\mathfrak{D}(\leq \mathbf{0}')$ *for each* $n \geq 3$.

(3) (Nies, Shore and Slaman, private communication) High$_n$, Low$_n$ *are definable in* $\mathfrak{C}$ *for* $n \geq 2, 3$ *(respectively).*

Improvements on these results seem to require discontinuity derived from the sort of local degree theory involving higher priority, in particular Lachlan [1975] and Harrington nonsplitting and $\emptyset'''$-priority. For example, splitting and relative nonsplitting techniques in $\mathfrak{D}(\leq \mathbf{0}')$ and the pseudo-jump machinery of Jockusch and Shore [1984] yield:

Theorem 3.3 (Cooper [ta1], [1994]). *There is a set P of degrees definable in $\mathfrak{D}$ which has greatest element $\mathbf{0}'$. Moreover, we can obtain $P = \mathfrak{C}$.*

Slaman and Woodin [1986] coding had previously led to the definability of $\mathfrak{C}$ in $\mathfrak{D}(\leq \mathbf{0}')$ from a finite number of parameters. The discontinuity here arises in showing the definability of uniformly low sets R, S of c.e. degrees which (by Welch [1981]) pairwise generate $\mathfrak{C}$. R, say, is defined by constructing $\mathbf{a}, \mathbf{b} < \mathbf{0}'$ such that

$$\mathbf{x} \in R \Leftrightarrow \mathbf{x} \text{ is a minimal solution of } \mathbf{x} \neq (\mathbf{x} \cup \mathbf{a}) \cap (\mathbf{x} \cup \mathbf{b}).$$

The above results seem close to the limits of what is possible with currently known techniques. Definability is dependent upon structural discontinuity, concerning which there are a number of intractible questions, which may turn out to be crucial.

4. Discontinuity in cones

Despite Theorem 2.2 above, the following discontinuity result is obtainable via an ingenious application of distributivity (Lachlan [1972]) of $\mathfrak{D}_{wtt}$ (the upper semilattice of c.e. wtt-degrees).

Theorem 4.1 (Stob [1983]). *There is a lower cone $\mathfrak{C}(\leq \mathbf{a})$ in which continuity of cupping fails.*

Proof. Construct a minimal pair $\mathbf{a}_0, \mathbf{a}_1$ of c.e. degrees whose join is *contiguous* (i.e. consists of just one wtt-degree).

To see that splitting of $\mathbf{a}_0 \cup \mathbf{a}_1$ is discontinuous at $\mathbf{a}_0, \mathbf{a}_1$, let $A_0, A_1 \in \mathbf{a}_0, \mathbf{a}_1$ respectively, and say $B \in \mathbf{b}$ with $\mathbf{b} \cup \mathbf{a}_1 = \mathbf{a}_0 \cup \mathbf{a}_1$. Then $A_0 \leq_{wtt} A_0 \oplus A_1 \leq_{wtt} B \oplus A_1$ (since $\mathbf{b} \cup \mathbf{a}_1$ is contiguous). Then by distributivity we have $A_0 \equiv_{wtt} B_0 \oplus A_{10}$, say, with $B_0 \leq_{wtt} B$, $A_{10} \leq_{wtt} A_1$. But $A_{10} \leq_{wtt} A_0, A_1 \Rightarrow A_{10}$ computable. So $A_0 \equiv_{wtt} B_0 \leq_{wtt} B$, giving $\mathbf{a}_0 \leq \mathbf{b}$. □

Although contiguous degrees are necessarily low$_2$, it can be shown that degrees $\mathbf{a}$ with discontinuous splittings are widely distributed in $\mathfrak{C}$ (see Cooper and Yi [1995]). For example, there exist such degrees $\mathbf{a}$ which are high.

For upper cones of $\mathfrak{C}$ there are some remarkable recent results.

We say that $\mathbf{a}$ *branches uniquely in* $\mathfrak{C}$ if and only if there exist c.e. $\mathbf{b}_0, \mathbf{b}_1 > \mathbf{a}$ for which

(i) $\mathbf{b}_0 \mid \mathbf{b}_1$,

(ii) $\mathbf{b}_0 \cap \mathbf{b}_1 = \mathbf{a}$, and

(iii) $\mathbf{c}_0 \cap \mathbf{c}_1 = \mathbf{a} \,\&\, \mathbf{c}_0 \mid \mathbf{c}_1 \Rightarrow (\exists j \leq 1)[\mathbf{c}_j \leq \mathbf{b}_0 \,\&\, \mathbf{c}_{1-j} \leq \mathbf{b}_1]$.

A particular case of a more general result of Ambos-Spies and Shore [1993] is:

Theorem 4.2 (Ambos-Spies and Ding [1994]). *There is a c.e.* $\mathbf{a}$ *which branches uniquely in* $\mathcal{C}$.

Proof. A combination of Lachlan [1966] branching and Fejer [1983] nonbranching.□

It follows from the proof (Shore, private communication) that the constructed branchings $\mathbf{b}_0, \mathbf{b}_1$ of $\mathbf{a}$ are already *individually* definable in $\mathcal{C}(\geq \mathbf{a})$.

One can also define a notion of *unique splitting* in intervals of $\mathcal{C}$. The following combines Sacks [1963] splitting with Harrington nonsplitting to extend a recent result of Leonhardi [1994].

Theorem 4.3 (Coles [ta]). *There is a c.e.* $\mathbf{a}$ *above which* $\mathbf{0}'$ *splits uniquely in* $\mathcal{C}$.

5. Definability in lower cones of $\mathcal{C}$

The following gives a special significance to definability in lower cones of $\mathcal{C}$.

Theorem 5.1 (Ambos-Spies, private communication). *Every lower cone* $\mathcal{C}(\leq \mathbf{a})$ *(*$\mathbf{a} \neq \mathbf{0}$*) is an automorphism base for* $\mathcal{C}$.

Proof. Let $\mathbf{a} > \mathbf{0}$, and let f be a nontrivial automorphism of $\mathcal{C}$ (with $\mathbf{a} = f(\mathbf{a})$, otherwise there is nothing to prove). Find a $\mathbf{c} \in \mathbf{PS}$ such that $\mathbf{c} \mid \mathbf{a}, f(\mathbf{c})$ and $\mathbf{c}$ has the '*downward separation property*' (that is, for all distinct $\mathbf{d}_0, \mathbf{d}_1 \mid \mathbf{c}$, $\mathcal{C}(< \mathbf{d}_0, \mathbf{c}) \neq \mathcal{C}(< \mathbf{d}_0, \mathbf{d}_1)$). Then $\mathcal{C}(< \mathbf{a}, \mathbf{c}) \neq \mathcal{C}(< \mathbf{a}, f(\mathbf{c})) = \mathcal{C}(< f(\mathbf{a}), f(\mathbf{c}))$. Since f is an automorphism there is some $\mathbf{d} \in \mathcal{C}(< \mathbf{a}, \mathbf{c})$ with $\mathbf{d} \neq f(\mathbf{d}) \in \mathcal{C}(< f(\mathbf{a}), f(\mathbf{c}))$. □

Corollary 5.2. *If there is a definable* $\mathbf{a} > \mathbf{0}$ *for which* $\mathcal{C}(\leq \mathbf{a})$ *is rigid, then* $\mathcal{D}$ *is rigid.*

Proof. By Slaman and Woodin [ta], if $\mathcal{C}$ is rigid so is $\mathcal{D}$. Similarly, the definability and rigidity of $\mathcal{C}(\leq \mathbf{a})$ gives the rigidity of $\mathcal{C}$, and hence of $\mathcal{D}$. □

A first step, both towards finding a rigid $\mathcal{C}(\leq \mathbf{a})$ *and* a definable $\mathbf{b} \in \mathcal{C}$, is to find a definable $\mathbf{b}$ in some lower cone $\mathcal{C}(\leq \mathbf{a})$.

Definition 5.3. $\mathbf{b}$ is a *nonsplitting base* in $\mathcal{C}(\leq \mathbf{a})$ if and only if $\mathbf{b} < \mathbf{a}$ and for all $\mathbf{c} \cup \mathbf{d} = \mathbf{a}$ *either* $\mathbf{c} \cup \mathbf{b} = \mathbf{a}$ *or* $\mathbf{d} \cup \mathbf{b} = \mathbf{a}$.

Theorem 5.4. *There is a lower cone* $\mathcal{C}(\leq \mathbf{c})$ *containing a least nonsplitting base* $\mathbf{a}$ *(so* $\mathbf{a}$ *is definable in* $\mathcal{C}(\leq \mathbf{c})$*).*

(We note that $\mathbf{c} \neq \mathbf{0}'$ by an unpublished result of Lachlan, Seetapun and Soare.)

Proof. We give a module for a typical batch of requirements. Let (W, U_0, U_1) and $(\Theta, \Phi, \Xi, \widehat{\Theta}_0, \widehat{\Theta}_1, \Psi)$ be typical entries in respective standard lists of all triples of c.e. sets and 6-tuples of p.c. functionals. We construct c.e. sets A, C, $B_{0,W,\Theta}$, $B_{1,W,\Theta}$ and p.c. functionals Δ, $\Omega_{W,\Theta}$, $\Upsilon_{0,W,\Theta}$, $\Upsilon_{1,W,\Theta}$, $\Gamma_{0,\Phi,W,\Theta}$, $\Gamma_{1,\Phi,W,\Theta}$, $\Lambda_{U,\Xi,\widehat{\Theta}}$, $\widehat{\Lambda}_{U,\Xi,\widehat{\Theta}}$ satisfying an overall requirement $A = \Delta^C$ and all requirements of the form:

$\mathcal{S}_{W,\Theta}$: $W = \Theta^C \Rightarrow C = \Omega_{W,\Theta}^{B_{0,W,\Theta},B_{1,W,\Theta}}$, $B_{i,W,\Theta} = \Upsilon_{i,W,\Theta}^C$ each $i \leq 1$,
(A requirement saying: If $\mathbf{w} \leq \mathbf{c}$ prepare a splitting $\mathbf{c} = \mathbf{b}_0 \cup \mathbf{b}_1$)

$\mathcal{P}_{i,\Phi,W,\Theta}$: $C = \Phi^{W,B_{i,W,\Theta}} \Rightarrow A = \Gamma_{i,\Phi,W,\Theta}^W$, $(i \leq 1)$,
(If $\mathbf{a} \not\leq \mathbf{w}$ then $\mathbf{w} \cup \mathbf{b}_0$, $\mathbf{w} \cup \mathbf{b}_1$ is a *nontrivial* splitting of $\mathbf{c}$)

$$\mathcal{Q}_{U,\Xi,\widehat{\Theta}} : \left.\begin{array}{l} C = \Xi^{U_0,U_1} \\ \& \, U_0 = \widehat{\Theta}_0^C \,\&\, U_1 = \widehat{\Theta}_1^C \end{array}\right\} \Rightarrow C = \Lambda_{U,\Xi,\widehat{\Theta}}^{U_0,A} \vee C = \widehat{\Lambda}_{U,\Xi,\widehat{\Theta}}^{U_1,A},$$

(Allow no nontrivial splittings of $\mathbf{c}$ above $\mathbf{a}$)

$\mathcal{R}_\Psi$: $C \neq \Psi^A$.

The atomic strategies.

(a) For $\mathcal{S}_{W,\Theta}$: We arrange all Θ-*expansionary* stages (that is, those stages at which the observed length of agreement $\ell(W, \Theta^C)$ of W and Θ^C attains a new maximum) to be $\Omega_{W,\Theta}$-expansionary stages. Following a C-change (at z say) we *rectify* $\Omega_{W,\Theta}$ (that is, redefine $\Omega_{W,\Theta}^{B_{0,W,\Theta},B_{1,W,\Theta}}(z) = C(z)$) using a $B_{i,W,\Theta} \upharpoonright \omega_{W,\Theta}(z)$-change, some $i = 0$ or 1 (permitted via $\Upsilon_{i,W,\Theta}$).

(b) For $\mathcal{P}_{i,\Phi,W,\Theta}$: All Φ-expansionary stages are arranged to be $\Gamma_{i,\Phi,W,\Theta}$-expansionary stages. Associate with each candidate w for membership of A an *agitator* a_w, and keep $\Gamma_{i,\Phi,W,\Theta}$ *honest* as far as possible by defining $\gamma(w) > \varphi(a_w)$. We respond to $w \searrow A$ by prompting $a_w \searrow C$ and imposing a $B_{i,W,\Theta}$-restraint (maintained until the next Φ- expansionary stage), using the consequent $W \upharpoonright \varphi(a_w)$-change to rectify $\Gamma_{i,\Phi,W,\Theta}$.

(c) For $\mathcal{Q}_{U,\Xi,\widehat{\Theta}}$: Application of the Lachlan [1975] nonsplitting strategy. That is, we use A-changes (below $\lambda(x)$ say) and C-restraints with the equations $U_i = \widehat{\Theta}_i^C$ to ensure to ensure that $\Lambda_{U,\Xi,\widehat{\Theta}}$ (and $\widehat{\Lambda}_{U,\Xi,\widehat{\Theta}}$) is honest at x. We rely on $x \searrow C$ producing a $U_0 \upharpoonright \xi(x)$- or $U_1 \upharpoonright \xi(x)$-change. We use a U_0-change to rectify $\Lambda_{U,\Xi,\widehat{\Theta}}$. We respond to a U_1-change with an $A \upharpoonright \lambda(x)$-change to rectify $\Lambda_{U,\Xi,\widehat{\Theta}}$, while implementing the auxiliary $\widehat{\Lambda}_{U,\Xi,\widehat{\Theta}}$-strategy.

As usual, the $\widehat{\Lambda}_{U,\Xi,\widehat{\Theta}}$-strategy will only in fact arise in response to unsuccessful attempts to satisfy a particular lower priority $\mathcal{R}_\Theta$-requirement, in which case the $\widehat{\Lambda}_{U,\Xi,\widehat{\Theta}}$-strategy is passed down to $\mathcal{R}_\Theta$-requirements of even lower priority in synchronisation with the $\Lambda_{U,\Xi,\widehat{\Theta}}$-activity in relation to the instigating $\mathcal{R}_\Theta$-requirement, and acts similarly to the $\Lambda_{U,\Xi,\widehat{\Theta}}$-strategy except in so far as it is assured (by astute choice of $\widehat{\Lambda}_{U,\Xi,\widehat{\Theta}}$-uses) of $\widehat{\Lambda}_{U,\Xi,\widehat{\Theta}}$-rectifying U_1-changes. Harrington honestification (to retain the proper relationship between Ξ- and $\Lambda_{U,\Xi,\widehat{\Theta}}$-,

$\widehat{\Lambda}_{U,\Xi,\widehat{\Theta}}$-uses) is avoided through controls on U_0, U_1 via the equations $U_0 = \widehat{\Theta}_0^C$, $U_1 = \widehat{\Theta}_1^C$.

(d) For $\mathcal{R}_\Psi$: A standard Friedberg–Mučnik strategy suffices.

There is a main conflict between these strategies which we address in the module below. In order to satisfy $\mathcal{R}_\Theta$, we may need to enumerate some x into C, while imposing an $A \upharpoonright \psi(x)$-restraint. *But* this may:

(i) lead to an $A \upharpoonright \lambda(x)$-change on behalf of $\mathcal{Q}_{\mathcal{U},\div,\widehat{\chi}}$ as a result of an instance of $\Lambda_{U,\Xi,\widehat{\Theta}}$-strategy failure (so injuring the A-restraint for $\mathcal{R}_\Theta$), and

(ii) force a $B_{i,W,\Theta}$-change for $\mathcal{S}_{\mathcal{W},\chi}$ which damages honesty in $\mathcal{P}_{\rangle,\oplus,\mathcal{W},\chi}$. *Then*

(iii) the A-change from (i) cannot be recorded by $\Gamma_{i,\Phi,W,\Theta}$ due to loss of honesty in (ii).

Basic module for $\mathcal{R}_\Psi$ below the $\Lambda_{U,\Xi,\widehat{\Theta}}$-strategy for $\mathcal{Q}_{U,\Xi,\widehat{\Theta}}$ operating below $\mathcal{S}_{W,\Theta}$ and $\mathcal{P}_{i,\Phi,W,\Theta}$.

Assume infinitely many Θ-, Φ- and Ξ-*expansionary stages*, organised in such a way that all Ξ-expansionary stages are Θ- and Φ-expansionary stages, and assume the module to only act at Ξ-expansionary stages.

(1) *Choose* a $\Gamma_{i,\Phi,W,\Theta}$*-rectification level* w for $\mathcal{R}_\Theta$.

(2) *Choose* a Φ*-agitator* $a_w > w$, and *define* $\delta(w) > a_w$.

(3) *Wait* for $\varphi(a_w)\downarrow$. Then *restrain* $B_{i,W,\Theta} \upharpoonright \varphi(a_w)$ and $C \upharpoonright \vartheta(\varphi(a_w))$ (both at $\mathcal{R}_\Theta$), and *define* $A = \Gamma^W_{i,\Phi,W,\Theta}$ up to a_w with $\gamma(w) > \varphi(a_w)$.

(4) *Choose* a $\Lambda_{U,\Xi,\widehat{\Theta}}$*-threshold* $t_w > \varphi(a_w)$ with $\Lambda^{U_0,A}_{U,\Xi,\widehat{\Theta}}(t_w)\uparrow$ and *wait* for $\Xi^{U_0,U_1}(t_w)\downarrow$.

(5) *Define* $\Lambda^{U_0,A}_{U,\Xi,\widehat{\Theta}}(t_w) = C(t_w)$ with $\xi(t_w) < \lambda(t_w)$ and $\Delta^C(\lambda(t_w))\uparrow$. *Restrain* $C \upharpoonright \widehat{\vartheta}_i(\xi(t_w))$, each $i \leq 1$, at $\mathcal{R}_\Theta$.

(6) *Choose* a potential witness $x > t_w$ (so $\omega(x) > \varphi(a_w)$) for $\mathcal{R}_\Theta$.

(7) *Wait* for $\Psi^A(x)\downarrow$.

(8) *Define* $C(x) \neq \Psi^A(x)$ and *restrain* $A \upharpoonright \psi(x)$. Then *ask:* Is $\lambda(t_w) > \psi(x)$?
  - **(a)** If *yes* - *Change* $B_{1-i} \upharpoonright \omega(x)$ and *proceed* to (9)(b). *Outcome:* $\mathcal{R}_\Theta$ satisfied.
  - **(b)** *No* – *Enumerate* t_w into C and *change* $B_{1-i} \upharpoonright \omega(t_w)$. *Proceed* to (9).

(9) *Ask:* Has $U_0 \upharpoonright \xi(t_w)$ changed?
  - **(a)** Yes – Use the U_0-change to *rectify* $\Lambda_{U,\Xi,\widehat{\Theta}}$. $\mathcal{R}_\Theta$ again satisfied.
  - **(b)** *No* - *Enumerate* $\lambda(t_w)$ into A and a_w into C. *Proceed* to (10), while *implementing* the auxiliary $\widehat{\Lambda}_{U,\Xi,\widehat{\Theta}}$-strategy.

(10) Use the $W \upharpoonright \varphi(t_w)$-change to *rectify* $\Gamma_{i,\Phi,W,\Theta}$ and then *return* to (1).

For further discussion of interactions along and off the true path see Cooper [ta2]

6. Definability in $\mathcal{D}(\leq \mathbf{a}\,\text{c.e.})$

Richer structure, often accessible via simpler techniques, is a feature of c.e. topped lower cones of Δ^0_2 degrees. There is an expectation that definability in such lower cones may be correspondingly more approachable, and that there may be an eventual dividend in the form of definability of c.e. singletons in $\mathcal{D}(\leq \mathbf{0}')$.

Bounding 1-generic degrees with non-zero c.e. **a**'s (see Odifreddi [ta], Jockusch [1980]) gives the rich basic structure for $\mathcal{D}(\leq \mathbf{a}\,\text{c.e.})$ (e.g., embeddings, minimal pairs, non-c.e. elements, not a lattice, etc) originally discovered by Sacks [1963] and Yates [1970]. The development of full-approximation tree constructions (Yates [1970] and Cooper [1973]) resulted in very strong initial segment embedding results for $\mathcal{D}(\leq \mathbf{a}\,\text{c.e.})$. Building on the Yates [1970] embedding of a minimal degree in any nontrivial c.e. topped lower cone, the most comprehensive such result is:

Theorem 6.1 (Lerman [1983]). *Every $\emptyset'$-presentable upper semi-lattice (with least element) is embeddable as an initial segment of $\mathcal{D}(\leq \mathbf{a}\,\text{c.e.})$ $(\mathbf{a} \neq \mathbf{0})$.*

(If $\mathbf{a} \leq \mathbf{0}'$ then $\mathcal{D}(\leq \mathbf{a})$ is $\mathbf{a}^{(3)} \leq \mathbf{0}^{(4)}$ presentable, so there is still considerable scope for improvement here. One should note however that any *lattice* embeddable as an initial segment of $\mathcal{D}(\leq \mathbf{a})$ must be Σ^0_3-presentable.)

Shore [1981] converts the structure provided by theorem 6.1 into definability.

Theorem 6.2 (Shore [1981]). *The theory of $\mathcal{D}(\leq \mathbf{a}\,\text{c.e.})$ $(\mathbf{a} \neq \mathbf{0})$ $\equiv_1$ the theory of first-order Peano arithmetic (so has degree $\mathbf{0}^{(\omega)}$). (But if $\mathcal{D}(\leq \mathbf{a}) \equiv \mathcal{D}(\leq \mathbf{0}')$ then $\mathbf{a}^{(3)} = \mathbf{0}^{(4)}$.)*

The most promising source of structure in the search for natural definitions of c.e. singletons in some $\mathcal{D}(\mathbf{0}, \mathbf{a}\,\text{c.e.})$ is:

Conjecture 6.3 (Epstein [1979]). *Every c.e. $\mathbf{b} \in \mathcal{D}(\mathbf{0}, \mathbf{a}\,\text{c.e.})$ has a minimal complement $\mathbf{m} < \mathbf{a}$.*

Epstein [1981] confirmed the conjecture for **a** high c.e., while Seetapun and Slaman [ta] have constructed a minimal complement for *any* $\mathbf{b} \in \mathcal{D}(\mathbf{0}, \mathbf{0}')$ (so extending the Posner [1981] Complementation Theorem). But for the general case, Cooper and Epstein [1987] pointed to a more interesting situation:

- If $\mathbf{b} < \mathbf{a}$ is low, then one can find a minimal $\mathbf{m} < \mathbf{a}$ with $\mathbf{m} \mid \mathbf{b}$. But there exists a $\mathcal{D}(\leq \mathbf{a}\,\text{c.e.})$ in which complementation fails (where by Cooper [1986], **a** can be chosen to be high).

  In fact:

- (Slaman, private communication) There exists a $\mathcal{D}(\leq \mathbf{a}\,\text{c.e.})$ in which *no* $\mathbf{b} \in \mathcal{D}(\mathbf{0}, \mathbf{a})$ is complemented.

- (Cooper [1989],Slaman and Steel [1988]) There exists a $\mathcal{D}(\leq \mathbf{a}\,\text{c.e.})$ in which cupping of some c.e. $\mathbf{b}$ fails. (For an interesting alternative proof of this see Downey [1987].) And:

- (Cooper [1986]) There exists a $\mathcal{D}(\leq \mathbf{a}\,\text{c.e.})$ in which capping of some $\mathbf{b}$ fails.

The above result of Cooper/ Slaman and Steel is the key to another, technically simpler, lower cone definition of a c.e. singleton.

Theorem 6.4. *There exist c.e. degrees* $\mathbf{0} < \mathbf{a} < \mathbf{c}$ *such that* $\mathbf{a}$ *is definable as the greatest degree in* $(\mathbf{0}, \mathbf{c})$ *not cuppable to* $\mathbf{c}$ *in* $\mathcal{D}(< \mathbf{c})$.

Proof. Let W and (Θ, Φ, Ξ, Ψ) be typical entries in respective standard lists of all c.e. sets and quadruples of p.c. functionals. We construct c.e. sets A, C, B_Ξ and p.c. functionals Ω, $\Gamma_{\Theta,\Phi}$, Λ_Ξ, Δ_Ξ, Υ_Ξ satisfying an overall requirement $A = \Omega^C$ and all requirements of the form:

$$\begin{aligned} \mathcal{P}_{\Theta,\Phi} &: \quad C = \Theta(A, \Phi^C) \Rightarrow A = \Gamma_{\Theta,\Phi}(\Phi^C), \\ \mathcal{Q}_W &: \quad A \neq \overline{W}, \end{aligned}$$

(Two typical requirements asking that $\mathbf{a}$ be noncuppable in $\mathcal{D}(\leq \mathbf{c})$)

$$\begin{aligned} \mathcal{S}_\Xi &: \quad C = \Lambda_\Xi(\Xi^C, B_\Xi) \text{ with } B_\Xi = \Delta_\Xi^C, \\ \mathcal{R}_{\Xi,\Psi} &: \quad C \neq \Psi^{B_\Xi} \vee \Xi^C = \Upsilon_\Xi^A \end{aligned}$$

(A pair of requirements asking that $\mathbf{a}$ be the largest such degree.)

Basic module for $\mathcal{P}_{\Theta,\Phi}$ above $\mathcal{Q}_W$ on the true path.

(1) *Choose* an *agitator* a for $\mathcal{P}_{\Theta,\Phi}$, $\mathcal{Q}_W$.
(2) *Wait* for $\ell(C, \Theta(A, \Phi^C)) > a$ and *restrain* $A \restriction \vartheta(a)$ and $C \restriction \varphi(\vartheta(a))$.
(3) *Choose* a witness x for $\mathcal{Q}_W$ with $\omega(x) > a$, $x > \vartheta(a)$, and $\Gamma_{\Theta,\Phi}(\Phi^C, x) \uparrow$ – *Define* $\Gamma_{\Theta,\Phi}(\Phi^C, x) = A(x)$ with $\gamma(x) > \vartheta(a)$.
(4) *Wait* for x realised (i.e., $x \in W$).
(5) *Enumerate* x into A and a into C (so rectifying Ω at x).
(6) Use the $\Phi^C \restriction \vartheta(a)$-change to rectify $\Gamma_{\Theta,\Phi}$ at x.

Module for $\mathcal{R}_{\Xi,\Psi}$ with $\mathcal{S}_\Xi$.

(1) *Choose* a witness y for $\mathcal{R}_{\Xi,\Psi}$ with $\Lambda_\Xi(\Xi^C, B_\Xi) \uparrow$.
(2) *Wait* for $\Xi^C \restriction \lambda(y) \downarrow$ and *restrain* it.
(3) *Wait* for y realised (i.e., $\ell(C, \Psi^{B_\Xi}) > y$).
(4) *Change* $C(y)$ and $B_\Xi(\lambda(y))$ (where $\delta(\lambda(y)) > y$), defining $\delta(\lambda(y)) > \xi(\lambda(y))$ whenever subsequently $\Xi^C(\lambda(y)) \downarrow$. *Restrain* $B_\Xi \restriction \psi(y)$ at all values $\neq \lambda(y)$.

(5) At each subsequent stage ask: Is $\Lambda_\Xi(\Xi^C, B_\Xi)\downarrow = 0$?
  (a) *Yes* – *Define* $B_\Xi(\lambda(y)) = 1$ and *initiate* an additional cycle of activity via (1).
  (b) *No* – *Define* $B_\Xi(\lambda(y)) = 0$ whenever $\Xi^C \upharpoonright \lambda(y)\downarrow$ with $\delta(\lambda(y)) > \xi(\lambda(y))$.

Infinitary outcome: We return to (1) infinitely often. In which case we apply the Modulus Lemma to obtain Ξ^C recursive.

Combining the strategies for the different groups of requirements

The main problem is that the activity of $\mathcal{Q}_W$ disrupts the application of the Modulus Lemma in the infinitary outcome for $\mathcal{R}_{\Xi,\Psi}$. However, all these C-injuries are A-related, so we are able to use these A-changes for rectification of Υ_Ξ with eventual outcome $\Xi^C = \Upsilon_\Xi^A$. Again, for a more detailed discussion see Cooper [ta2]. □

References

K. Ambos-Spies and D. Ding [1994], Discontinuity of cappings in the recursively enumerable degrees and strongly nonbranching degrees, Math. Logic Quarterly 40, 287–317.

K. Ambos-Spies and R. A. Shore [1993], Undecidability and 1-types in the recursively enumerable degrees, Ann. Pure Appl. Logic 63, 3–37.

K. Ambos-Spies, A. H. Lachlan and R. I. Soare [1993], The continuity of cupping to $\mathbf{0}'$, Ann. Pure Appl. Logic 64, 195–202.

M. M. Arslanov, S. Lempp and R. A. Shore [1996], On isolating r.e. and isolated d-r.e. degrees, Computability, Enumerability, Unsolvability: Directions in Recursion Theory (S. B. Cooper, T. A. Slaman and S. S. Wainer, eds.), London Math. Soc. Lecture Notes, Cambridge University Press, Cambridge.

Calhoun and T. A. Slaman [ta], The Π_2^0 e-degrees are not dense, to appear in the J. of Symbolic Logic.

R. Coles [ta], An upper cone with a unique Harrington splitting, in preparation.

A. Church [1936], A note on the Entscheidungsproblem, J. Symbolic Logic 1, 40–41 and 101–102.

S. B. Cooper [1973], Minimal degrees and the jump operator, J. Symbolic Logic 38, 249–271.

S. B. Cooper [1989], The strong anti-cupping property for recursively enumerable degrees, J. Symbolic Logic 54, 527–539.

S. B. Cooper [1986], Some negative results on minimal degrees below $\mathbf{0}'$, (abstract), Recursive Function Theory Newsletter, Item 353, Issue 34, January 1986.

S. B. Cooper [1990], Enumeration reducibility, nondeterministic computations and relative computability of partial functions, in: Recursion Theory Week, Oberwolfach 1989 (K. Ambos-Spies, G. Müller, G. E. Sacks, eds.), Lecture Notes in Math. 1432, Springer-Verlag, Berlin, Heidelberg, New York, pp. 57–110.

S. B. Cooper [1994], Rigidity and definability in the non-computable universe, in: Logic, Methodology and Philosophy of Science IX, Uppsala, Sweden, August 1991 (D. Prawitz, B. Skyrms and D. Westerstahl, eds.), Stud. Logic Found. Math. 134, North-Holland, Amsterdam, pp. 209–236.

S. B. Cooper [ta1], On a conjecture of Kleene and Post, to appear.

S. B. Cooper [ta2], Definability in intervals of the computably enumerable degrees, in preparation.

S. B. Cooper and R. L. Epstein [1987], Complementing below recursively enumerable degrees, Ann. Pure Appl. Logic 34, 15–32.

S. B. Cooper, L. Harrington, A. H. Lachlan, S. Lempp and R. I. Soare [1991], The D.R.E. degrees are not dense, Ann. Pure Appl. Logic 55, 125–151.

S. B. Cooper and X. Yi [1995], Discontinuity of splitting in the recursively enumerable degrees, Archive for Math. Logic 34, 247–256.

R. G. Downey [1987], Δ_2^0 degrees and transfer theorems, Illinois J. Math. 31, 419–427.

R. L. Epstein [1979], Degrees of Unsolvability: Structure and Theory, Lecture Notes in Math. 759, Springer-Verlag, Berlin, Heidelberg, New York.

R. L. Epstein [1981], Initial segments of degrees below $\mathbf{0}'$, Memoirs Amer. Math. Soc. 241, Amer. Math. Soc., Providence, R.I.

S. Feferman [1957], Degrees of unsolvability associated with classes of formalized theories, J. Symbolic Logic 22, 161–175.

P. A. Fejer [1983], The density of the nonbranching degrees, Ann. Pure Appl. Logic 24, 113–130.

R. M. Friedberg [1957], Two recursively enumerable sets of incomparable degrees of unsolvability, Proc. Natl. Acad. Sci. USA 43, 236–238.

K. Gödel [1931], Über formal unentscheidbare Sätze der Principia Mathematica und verwandter Systeme I, Monatsh. Math. Phys. 38, 173–198.

K. Gödel [1934], On undecidable propositions of formal mathematical systems, (mimeographed notes), The Undecidable. Basic Papers on Undecidable Propositions, Unsolvable Problems, and Computable Functions (M. Davis, ed.), Raven Press, New York, 1965, pp. 39–71.

W. Hanf [1965], Model theoretic methods in the study of elementary logic, in: Symposium on the Theory of Models (J. W. Addison, L. Henkin and A. Tarski, eds.) North-Holland, Amsterdam, pp. 132–145.

L. Harrington and R. I. Soare [1992], Games in recursion theory and continuity properties of capping degrees, in: Set Theory of the Continuum (H. Judah, W. Just, and W. H. Woodin, eds.) Math. Sci. Res. Inst. Publ. 26, Springer-Verlag, Berlin, Heidelberg, New York, pp. 39–62.

L. Harrington and R. I. Soare [1991], Post's program and incomplete recursively enumerable sets, Proc. Natl. Acad. Sci. USA 88, 10242–10246.

C. G. Jockusch, Jr. [1980], Degrees of generic sets, in: Recursion Theory: Its Generalisations and Applications (F. Drake and S. S. Wainer, eds.) Proceedings of Logic

Colloquium '79, London Math. Soc. Lecture Note Ser. 45, Cambridge University Press, Cambridge, pp. 110–139.

C. G. Jockusch, Jr. and R. A. Shore [1984], Pseudo jump operators II: Transfinite iterations, hierarchies, and minimal covers, J. Symbolic Logic 49 , 1205–1236.

C. G. Jockusch, Jr. and S. G. Simpson [1976], A degree theoretic definition of the ramified analytical hierarchy, Ann. Math. Logic 10, 1–32.

A. H. Lachlan [1966], Lower bounds for pairs of recursively enumerable degrees, Proc. London Math. Soc. 16, 537–569.

A. H. Lachlan [1972], Recursively enumerable many-one degrees, Algebra i Logika 11, 326–358 (Russian); Algebra and Logic 11, 186–202 (English translation).

A. H. Lachlan [1975], A recursively enumerable degree which will not split over all lesser ones, Ann. Math. Logic 9, 307–365.

A. H. Lachlan [1975a], Uniform enumeration operators, J. Symbolic Logic 40, 401–409.

A. H. Lachlan [1979], Bounding minimal pairs, J. Symbolic Logic 44, 626–642.

S. Leonhardi [1994], Generalised Nonsplitting in the Recursively Enumerable Degrees, Ph.D. Dissertation, University of Wisconsin-Madison, August 1994.

M. Lerman [1983], Degrees of Unsolvability, Perspectives in Mathematical Logic, Omega Series, Springer-Verlag, Berlin, Heidelberg, London, New York, Tokyo.

S. S. Marchenkov [1976], A class of incomplete sets , Mat. Zametki 20, 473–478 (Russian); Math. Notes 20 (1976), 823–825 (English translation).

Ju. V. Matijasevič [1970], Enumerable sets are diophantine, Dokl. Akad. Nauk. SSSR 191, 279–282 (Russian); Sov. Math. Dokl. 11, 354–357 (English translation).

A. A. Mučnik [1956], On the unsolvability of the problem of reducibility in the theory of algorithms, Dokl. Akad. Nauk SSSR 108, 194–197 (Russian).

J. Myhill [1955], Creative sets, Z. Math. Logik Grundlag. Math. 1, 97–108.

A. Nerode and R. A. Shore [1980], Second order logic and first order theories of reducibility orderings, in: The Kleene Symposium (J. Barwise, H. J. Keisler and K. Kunen, eds.), North-Holland, Amsterdam, pp. 181–200.

A. Nerode and R. A. Shore [1980a], Reducibility orderings: theories, definability and automorphisms, Ann. Math. Logic 18, 61–89.

P. Odifreddi [1989], Classical Recursion Theory, North-Holland, Amsterdam, New York, Oxford.

P. Odifreddi [ta], Classical Recursion Theory, Vol. II, North-Holland, Amsterdam, New York, Oxford, in preparation.

D. Posner [1981], The upper semi-lattice of degrees $\leq \mathbf{0}'$ is complemented, J. Symbolic Logic 46, 705–713.

E. L. Post [1944], Recursively enumerable sets of positive integers and their decision problems, Bull. Amer. Math. Soc. 50 (1944), 284–316.

R. W. Robinson [1971], Interpolation and embedding in the recursively enumerable degrees, Ann. of Math. (2) 93, 285–314.

G. E. Sacks [1963], Degrees of Unsolvability, Ann. of Math. Stud. 55, Princeton University Press, Princeton, N.J.

G. E. Sacks [1964], The recursively enumerable degrees are dense, Ann. of Math. (2) 80, 300–312.

G. E. Sacks [1967], On a theorem of Lachlan and Martin, Proc. Amer. Math. Soc. 18, 140–141.

G. E. Sacks [1985], Some open questions in recursion theory, in: Recursion Theory Week (H. D. Ebbinghaus, G. H. Müller and G. E. Sacks, eds.), Lecture Notes in Mathematics No. 1141, Springer-Verlag, Berlin, Heidelberg, New York, pp. 333–342.

D. Seetapun [ta], Every recursively enumerable degree is locally noncappable, to appear.

D. Seetapun and T. A. Slaman [ta], Minimal complements, to appear.

J. R. Shoenfield [1958], Degrees of formal systems, J. Symbolic Logic 23, 389–392.

J. R. Shoenfield [1960], Degrees of models, J. Symbolic Logic 25, 233–237.

R. A. Shore [1981], The theory of the degrees below $\mathbf{0}'$, J. London Math. Soc. 24, 1–14.

R. A. Shore [1988], Defining jump classes in the degrees below $\mathbf{0}'$, Proc. Amer. Math. Soc. 104, 287–292.

S. G. Simpson [1977], First-order theory of the degrees of recursive unsolvability, Ann. of Math. (2) 105, 121–139.

T. A. Slaman [1991], Degree structures, Proc. Int. Congress of Math., Kyoto, 1990, Springer-Verlag, Tokyo, pp. 303–316.

T. A. Slaman [ta], Questions in Recursion Theory, limited circulation, July 1994.

T. A. Slaman and J. R. Steel [1988], Definable functions on degrees, in: Cabal Seminar 81–85 (eds. A. S. Kechris, D. A. Martin and J. R. Steel), Lecture Notes in Math. 1333, Springer-Verlag, Berlin, pp. 37–55.

T. A. Slaman and J. R. Steel [1989], Complementation in the Turing degrees, J. Symbolic Logic 54, 160–176.

T. A. Slaman and W. H. Woodin [1986], Definability in the Turing degrees, Illinois J. Math. 30, 320–334.

T. A. Slaman and W. H. Woodin [ta], Definability in degree structures, to appear.

R. I. Soare [1987], Recursively enumerable sets and degrees, Springer-Verlag, Berlin, Heidelberg, London, New York.

M. Stob [1983], wtt-degrees and T-degrees of recursively enumerable sets, J. Symbolic Logic 48, 921–930.

A. M. Turing [1936], On computable numbers, with an application to the Entscheidungsproblem, Proc. London Math. Soc. 42, 230–265.

L. V. Welch [1981], A Hierarchy of Families of Recursively Enumerable Degrees and a Theorem on Bounding Minimal Pairs, Ph.D. Dissertation, University of Illinois at Urbana-Champaign.

C. E. M. Yates [1970], Initial segments of the degrees of unsolvability, Part II: Minimal degrees, J. Symbolic Logic 35, 243–266.

Iterated skew polynomial rings and simple modules

David A. Jordan

Introduction

There are many instances of skew polynomial algebras in two indeterminates x and y over an algebra A where, for some scalar ρ, the "skew" commutator, $xy - \rho yx$, is in A. The most familiar examples are probably the Weyl algebra A_1, where A is a field, and $xy - yx = 1$ and the enveloping algebra of the Lie algebra sl_2, where A is a polynomial algebra in one variable, (which is usually written h with e and f for x and y). Recently many new examples of such algebras have emerged, usually with the word "quantum" in their name. In a sequence of papers [9, 10, 11, 12, 13, 14, 15], we have developed the theory of these rings with a view to explaining the similarities and the differences which exist between the motivating examples. In [9] we described a construction which yielded some of these rings with a commutative base ring and with the scalar $\rho = 1$. To accommodate more examples, this was later extended to include arbitrary values of ρ in [11, 14] and non-commutative base rings in [15]. A restriction on the element $xy - \rho yx$ of A was removed by Wells in [23] in order to include some examples of simple rings.

Sections 1 and 3 are mostly expository but include some short proofs which give a flavour of proofs of more general results published elsewhere. Section 2 contains new results on simple modules for the rings R given by the adapted construction. An analysis of the finite-dimensional simple modules for the case of a commutative base ring A is given in [14]. Essentially, there are two types of modules each of which extend simple A-modules in a natural way. These modules are often simple but when they are not they have factors which are finite-dimensional. We consider the generalizations of these two types of module to the case of a noncommutative base ring and discuss a number of examples of simple modules. We also comment on the existence of simple modules of different Gelfand–Kirillov dimension for the simple algebras considered in [18].

I would like to thank the Conference organizers, and in particular Professor S. Skryabin, for the invitation to Kazan and for their hospitality.

1. Construction and examples

1.1. The general construction

Let k be a field and let A be a k-algebra with two commuting k-automorphisms, α and β. Let $\gamma = \alpha\beta$ and suppose that γ is normalizing, that is there exists a regular element $r \in A$ such that $ra = \gamma(a)r$ for all $a \in A$. Let v be a normal element, not necessarily regular, of A such that $va = \gamma(a)v$ for all $a \in A$. Finally, let $\rho \in k^*$, the multiplicative group of k. From this data we constuct a k-algebra R as follows: R is the iterated skew polynomial ring $A[y;\alpha][x;\beta,\delta]$ where $\beta(y) = \rho y, \delta(A) = 0$ and $\delta(y) = v$. Thus

$$ya = \alpha(a)y; \quad xa = \beta(a)x; \quad xy - \rho yx = v.$$

Here δ is an β-derivation,

$$\delta(rs) = \beta(r)\delta(s) + \delta(r)s \; \forall r, s \in A[y;\alpha].$$

1.2. Examples with A commutative

When the base ring A is commutative, the automorphism γ is necessarily the identity automorphism id_A and $\beta = \alpha^{-1}$. Also v may be any element of A.

(i) With $A = k, \alpha = \mathrm{id}, v = 1$ and $\rho = q$, the k-algebra R is the k-algebra generated by x and y subject to the single relation $xy - qyx = 1$. This is the quantized Weyl algebra A_1^q and in the case $q = 1$ is the usual Weyl algebra A_1.

(ii) With $A = k[t], \alpha(t) = t + 2, u = -\frac{1}{4}(t-1)^2$ and $\rho = 1$, the k-algebra R is generated by t, x and y subject to the relations

$$tx - xt = 2x, \quad yt - ty = 2y, \quad xy - yx = t.$$

This is the enveloping algebra of the Lie algebra sl_2.

(iii) With $A = k[t]$, $\alpha(t) = t + 1$, $\rho = -1$ and $v = t$, R is generated by t, x and y subject to the relations

$$tx - xt = x, \quad yt - ty = y, \quad xy + yx = t.$$

When $k = \mathbb{C}$, this is the universal enveloping algebra of the dispin Lie superalgebra $B[0,1]$, see [17, 21].

(iv) Let $A = k[t]$, let $\nu \in k\backslash\{0, 1, -1\}$, let α be the k-automorphism of A such that $\alpha(t) = \nu^4 t$, let $\rho = \nu^{-2}$ and let $v = \nu^{-1}(t + \frac{\nu^2}{\nu^2-1})$. Then, as a k-algebra, R is generated by t, x and y subject to the relations

$$xt = \nu^{-4}tx, \quad yt = \nu^4 ty, \quad xy - \nu^{-2}yx = \nu^{-1}(t + \frac{\nu^2}{\nu^2 - 1}).$$

With $t' = t + \frac{\nu^2}{\nu^2-1}$, these become

$$xt' - \nu^{-4}t'x = (1+\nu^{-2})x, \quad t'y - \nu^{-4}yt' = (1+\nu^{-2})y, \quad xy - \nu^{-2}yx = \nu^{-1}t'.$$

This algebra was first considered by Woronowicz [24], see also [21, §7].

If, in this example, $k[t]$ is replaced by the Laurent polynomial algebra $k[t, t^{-1}]$ and if ν is not a root of unity then A now has no non-zero ideal invariant under α. The resulting ring R differs from the original in a number of respects. For example, it has a non-trivial centre and the localization which converts the original ring R annihilates an infinite family of 1-dimensional simple modules, see [14].

(v) If A is the commutative polynomial ring $k[b, c]$, $\alpha(b) = q^{-2}b, \alpha(c) = q^{-2}c, u = q^2bc$, $\rho = 1$ and $x = a, y = d$ then the algebra R is one version of the coordinate ring $\mathcal{O}_q(M_2(k))$ of quantum 2×2 matrices. A more general version will be discussed in 1.4.

1.3. Normal elements

One common feature of the algebras in 1.2 is that, with the exception of the case $q = 1$ in (i), that is the case of the Weyl algebra A_1, each has a distinguished normal element which is significant in the study of the algebra. Examples are the Casimir element Ω in (ii) and the quantum determinant in (v). In those cases where $\rho = 1$, this normal element is central. These normal elements have the same basic form and there is a condition on v which guarantees their existence.

Suppose that $v = u - \rho\alpha(u)$ for some normal element $u \in A$ which also induces γ. As in [16], we shall refer to the case where v has this form as the *conformal* case. In 1.2(i), v has this form if and only if $q \neq 1$, in which case $u = \frac{1}{1-q}$.

In the conformal case, the element $z = xy - u = \rho(yx - \alpha(u))$ satisfies the following identities:

$$zy = \rho yz, \quad zx = \rho^{-1}xz, \quad za = \gamma(a)z \forall a \in A.$$

Consequently z is a normal element of R inducing a k-automorphism σ of R such that $\sigma(a) = a$ for all $a \in A$, $\sigma(y) = \rho y$ and $\sigma(x) = \rho^{-1}x$. Note that if $\rho^n = 1$ and $\gamma^n = \mathrm{id}_A$ then z^n is central. In particular if $\rho = 1$ and $\gamma = \mathrm{id}_A$ then z is central.

To match the existing literature on a particular algebra, it is often necessary to replace z by an element of the form $\lambda z + \mu$ where λ and μ are scalars with $\lambda \neq 0$ and, unless $\rho = 1$, $\mu = 0$. For example, in the quantized Weyl algebra in 1.2(i), if $q \neq 1$ the normal element $z = xy - \frac{1}{1-q} = q(yx - \frac{1}{1-q})$ is usually replaced by $\frac{q-1}{q}z = 1 + (q-1)yx = xy - yx$. In 1.2(ii) the central element $z = \frac{1}{4}(\Omega + 1)$, where Ω is the Casimir element of $U(sl_2)$. We shall sometimes use the name Casimir element for $z = xy - u$ in the conformal case.

The normal element z, when it exists, is never a unit. Consequently the non-existence of an element u as above is a necessary condition for the simplicity of R. Simplicity is discussed in detail in [23] where necessary and sufficient conditions for simplicity are established for R and for the rings which can be obtained, in the conformal case, from R by localization at the powers of z or by factoring out the ideal zR. The basic motivating example for studying simplicity in this context is the quantized Weyl algebra, 1.2(i), which is simple if and only if $q = 1$ and k has characteristic 0.

1.4. Higher quantized Weyl algebras

The existence of the normal element z in the conformal case allows for the construction in 1.1 to be iterated with R replacing A and a scalar multiple of z in the role of v.

Let $A = A_1^q$ with x, y and q renamed as x_1, y_1 and q_1 respectively. Then $z_1 = 1 + (q_1 - 1)y_1x_1$ is normal. Given $0 \neq \lambda \in k$, there is a k-automorphism α of A such that

$$\alpha(y_1) = \lambda y_1; \quad \alpha(x_1) = \lambda^{-1}x_1.$$

Let $\rho = q_2 \in k\backslash\{0,1\}$, let $u = \frac{1}{1-q_2}$ and form $R = R(A, \alpha, u, \rho)$. This gives the k-algebra generated by x_1, x_2, y_1 and y_2 subject to the relations

$$\begin{aligned}
x_1y_1 - q_1y_1x_1 &= 1; \\
y_2y_1 &= \lambda y_1y_2; \\
\lambda y_2x_1 &= x_1y_2; \\
\mu x_2x_1 &= x_1x_2; \\
x_2y_1 &= \mu y_1x_2; \\
x_2y_2 - q_2y_2x_2 &= 1 + (q_1 - 1)y_1x_1,
\end{aligned}$$

where $\mu = q_1\lambda^{-1}$. This is the second quantized Weyl algebra $A_2^{\overline{q},\lambda}$. Inductively, the nth quantized Weyl algebra $A_n^{\overline{q},\Lambda}$, where $\overline{q} = (q_1, q_2, \ldots, q_n)$ and Λ is a triangular array of parameters λ_{ij}, is constructed from $A_{n-1}^{\overline{q},\Lambda}$ in the same way. It has a sequence of n normal elements $z_1, z_2, \ldots, z_n$ which commute with each other. They are scalar multiples of the Casimir elements which arise at each step and can be written in the form $z_i = x_iy_i - y_ix_i$. Full details may be found in [15] where the following is proved.

1.5. Theorem. *If no q_i is a root of unity then the localization of $A_n^{\overline{q},\Lambda}$ at the multiplicatively closed set $\mathcal{Z}$ generated by the z_i's is a simple Noetherian ring.*

This theorem indicates that the localization is a more appropriate analogue of the ordinary Weyl algebra A_n than $A_n^{\overline{q},\Lambda}$ itself.

1.6. Further examples with *A* noncommutative

Starting with the field k and applying the construction 1.1 twice with a non-zero initial choice of v, followed by a non-zero scalar multiple of z (in the conformal case) or 1 gives a ring isomorphic to the second quantized Weyl algebra. Some different significant examples can be obtained from k by two applications of 1.1 with the initial choice $v = 0$. The ring obtained after one application of 1.1 is then the k-algebra generated by y_1 and x_1 subject to the relation

$$x_1y_1 = \rho_1y_1x_1.$$

This is the coordinate ring of the quantum plane and we shall denote it by A with a view to applying 1.1 a second time. The normal element z here is $z_1 = x_1y_1$. Let $a, b \in k\backslash\{0\}$. Then there is a k-automorphism α of A with $\alpha(x) = ax_1$ and $\alpha(y_1) = by_1$. Let $\tau \in k\backslash\{0\}$ and let $u = \tau z_1$. Applying 1.1 to A, α and v where $\rho = \rho_2$ and $v = u - \rho\alpha(u)$, gives the k-algebra R generated by x_1, x_2, y_1 and y_2 subject to the relations:

$$\begin{aligned} x_1y_1 &= \rho_1y_1x_1; \\ y_2x_1 &= ax_1y_2; \\ y_2y_1 &= by_2y_1; \\ \rho_1ax_2x_1 &= x_1x_2; \\ bx_2y_1 &= \rho_1y_1x_2; \\ x_2y_2 - \rho_2y_2x_2 &= \tau(1-\rho_2ab)x_1y_1. \end{aligned}$$

If q and p are non-zero scalars and we set $a = pq, b = p^{-1}, \rho_1 = (p^2q)^{-1}, \rho_2 = 1$ and $\tau = p$ then these relations, with appropriate renaming of the indeterminates, are those given in [1] for a two-parameter version of the coordinate ring $\mathcal{O}_q(M_2(k))$ of quantum 2×2 matrices. This includes the one-parameter quantizations given in [6], where $\rho_1 = 1$ and the intermediate ring is commutative, and in [5].

These examples are conformal and the normal element z of 1.3 is the so-called *quantum determinant*. The coordinate ring of the quantum group $\mathcal{O}_q(GL_2(k))$ is obtained by localization at the powers of z and, when z is central, $\mathcal{O}_q(SL_2(k))$ is obtained by factoring out the ideal $(z-1)R$.

In [22], Vancliffe studies a class of quadratic algebras which splits into two subclasses. If α, β and q are scalars with $\alpha, q \neq 0$ and $\alpha\beta \neq \lambda q$ then setting $a = \alpha, b = q^{-1}, \rho_1 = (\alpha q)^{-1}, \rho_2 = \beta$ and $\tau = -q$ above gives rise, with appropriate renaming of the indeterminates, to the first of these. As observed in [22, 1.7], this includes the coordinate ring of quantum symplectic 4-dimensional space from [6]. Further iteration of the construction yields the coordinate rings of higher-dimensional quantum symplectic spaces for which the primitive ideals were analysed by Oh in [19].

The second of the subclasses from [22] can also be constructed from k with two applications of 1.1 but this time the second choice for u is not a scalar multiple of the first Casimir element. As above the initial choice of u is 0 but ρ_1 is taken to be -1, giving the k-algebra A generated by x_1 and y_1 subject to the relation $x_1y_1 + y_1x_1 = 0$. The construction in 1.1 is then applied to A with $\alpha(x_1) = \lambda x_1$, $\alpha(y_1) = -\lambda y_1$, $\rho\lambda^2 \neq 1$, where $\lambda, \rho \in k^*$, and $v = u - \rho\alpha(u)$, where $u = \dfrac{x_1^2 + y_1^2}{1 - \rho\lambda^2}$ which is central in A.

1.7. Skew commutator formulae

Below we list some formulae for skew commutators involving y or x and a power of the other. These are readily checked inductively and are extremely useful in calculations involving the rings R.

For $n \geq 1$, let $v_n = \sum_{j=0}^{n-1} \rho^j \alpha^j(v)$, a normal element of A such that $v_n a = \gamma(a) v_n \ \forall a \in A$.

(i) $xy^n - \rho^n y^n x = v_n y^{n-1}$.

(ii) $x^n y - \rho^n y x^n = \alpha^{1-n}(v_n) x^{n-1}$.

In the conformal case, $v_n = (u - \rho^n \alpha^n(u))$ so (i) and (ii) become

(iii) $xy^n - \rho^n y^n x = (u - \rho^n \alpha^n(u)) y^{n-1}$ and

(iv) $x^n y - \rho^n y x^n = (\alpha^{1-n}(u) - \rho^n \alpha(u)) x^{n-1}$.

In A_1^q with $q \neq 1$ (iii) becomes

(v) $xy^n - q^n y^n x = \dfrac{1-q^n}{1-q} y^{n-1}$.

As an example of the use of these formulae in "degree reduction" arguments, we give the proof of 1.1 for the case $n = 1$. Letting S be the localization of A_1^q at the powers of z, the localization S_y at the powers of y is $S_y = k[z^{\pm 1}][y^{\pm 1}; \gamma]$ with $\gamma(z) = q^{-1} z$. When q is not a root of unity, this is well-known to be simple. Consequently any non-zero ideal I of S contains a power of y and by repeated use of (v), $1 \in I$. Thus S is simple when q is not a root of unity.

Formula (v) also enables us to view x in A_1^q as a difference operator on $k[y]$. Let $R = A_1^q$. Then, as a k-vector space, the left R-module $\frac{R}{Rx}$ is isomorphic to $k[y]$ and this gives $k[y]$ a left R-module structure where

$$x.y^n = \frac{q^n - 1}{q - 1} y^{n-1},$$

$$x.f(y) = \frac{f(qy) - f(y)}{qy - y}.$$

Thus x acts as the so-called *q-difference operator* (or *Euler* derivative or *Jackson* derivative).

These ideas extend to higher quantized Weyl algebras. For example, the actions of the four generators x_1, x_2, y_1 and y_2 of $R = A_2^{\bar{q},\lambda}$ on a typical monomial $y_1^i y_2^j$ in the left R-module $V = k[y_1, y_2]$, which is identified with $\dfrac{R}{Rx_1 + Rx_2}$, are given below. Notice that, for $i, j = 1, 2$, with $i \neq j$, x_i acts as the q_i-difference operator on $k[y_i]$ and as 0 on $k[y_j]$.

$$\begin{aligned}
x_1 : y_1^i y_2^j &\mapsto \frac{q_1^i - 1}{q_1 - 1} y_1^{i-1} y_2^j; \\
x_2 : y_1^i y_2^j &\mapsto \mu^i \frac{q_2^j - 1}{q_2 - 1} y_1^i y_2^{j-1}; \\
y_1 : y_1^i y_2^j &\mapsto y_1^{i+1} y_2^j;
\end{aligned}$$

$$y_2 : y_1^i y_2^j \quad \mapsto \quad \lambda^i y_1^i y_2^{j+1}.$$

It is an instructive exercise to check that these four linear transformations of V satisfy the six defining relations for R and that, provided q_1 and q_2 are not roots of unity, R can be defined as the subalgebra of $\mathrm{End}_k V$ that they generate.

The normal elements z_1, z_2 act on V as automorphisms:

$$\begin{aligned} z_1 : y_1^i y_2^j \quad &\mapsto \quad q_1^i y_1^i y_2^j; \\ z_2 : y_1^i y_2^j \quad &\mapsto \quad q_1^i q_2^j y_1^i y_2^j. \end{aligned}$$

One result of localizing at $\{z_1^i z_2^j\}$ is to introduce more symmetry to the situation through the action of $z_2 z_1^{-1}$ which mirrors that of z_1. Another is that the localization contains a copy of $A_1^{q_2}$ as well as the obvious copy of $A_1^{q_1}$. Postmultiplication of the formula $x_2 y_2 - y_2 x_2 = z_2$ by z_2^{-1} and use of the relation $x_2 z_2^{-1} = q_2^{-1} z_2^{-1} x_2$ give $x_2' y_2' - q_2 y_2' x_2' = 1$, where $x_2' = y_2 z_2^{-1}$ and $y_2' = -q_2^{-1} x_2$.

The above remarks on symmetry and partial difference operators generalize routinely from 2 to n.

1.8. Localizations and Grading

Let R be as in 1.1 and let $w = xy$. Then $wa = \gamma(a)w$ for all $a \in A$ and, in the conformal case, $w = z + u$. As the set $\mathcal{Y} = \{y^i\}_{i \geq 1}$ is a β-invariant right and left Ore set in $A[y;\alpha]$ so, by [7, Lemma 1.4], it is a right and left Ore set in R. As $x = (z+u)y^{-1}$, the localization $R_{\mathcal{Y}}$ is generated, as a ring extension of A, by y, y^{-1} and w, or, in the conformal case, by y, y^{-1} and z. We say that a left R-module is y-torsion (resp. y-torsion-free) if it is torsion (resp. torsion-free) with respect to $\mathcal{Y}$.

The ring R has a $\mathbb{Z}$-grading in which y and x are homogeneous of degrees 1 and -1 respectively and each element a of A has degree 0. Using this grading, it can be checked that, as an A-module, $R_{\mathcal{Y}}$ is freely generated by the set $\{w^i y^j : i \geq 0, j \in \mathbb{Z}\}$ and hence that $R_{\mathcal{Y}}$ can be identified with $A[w;\gamma][y, y^{-1};\alpha]$, where α extends to $A[w;\gamma]$ with $\alpha(w) = \rho^{-1}(w - v)$. In the conformal case, w can be replaced by the normal element z which satisfies $\alpha(z) = \rho^{-1}z$. The identities (i) and (ii) below, and their alternatives (iii) and (iv) for the conformal case, are easily checked inductively and show that the degree 0 component R_0 is the subring $A[w;\gamma]$ (or, in the conformal case, $A[z;\gamma]$) and that, for $n \neq 0$, the component R_n is $R_0 y^n$ if $n > 0$ and $R_0 x^{-n}$ if $n < 0$.

(i) $x^n y^n = \prod_{j=0}^{n-1} \alpha^{-j}(w)$.

(ii) $y^n x^n = \prod_{j=1}^{n} \alpha^j(w)$.

(iii) $x^n y^n = \rho^{\frac{n(n-1)}{2}}(z + \rho^{1-n}\alpha^{1-n}(u)) \ldots (z + \rho^{-1}\alpha^{-1}(u))(z+u)$.

(iv) $y^n x^n = \rho^{-\frac{n(n+1)}{2}}(z + \rho^n \alpha^n(u)) \ldots (z + \rho\alpha(u))(z + \alpha(u))$.

In the construction described in 1.1, there is symmetry in the roles of y and x. If these are interchanged, and u, ρ and α are replaced by $-\alpha(u)$, ρ^{-1} and β

respectively then the same ring results. Thus, for example, $\mathcal{X} = \{x^i\}_{i\geq 1}$ is an Ore set in R with $R_{\mathcal{X}} = A[w;\gamma][x, x^{-1}; \beta]$.

1.9. Factoring out the Casimir element

In the conformal case, the factor ring $\frac{R}{zR}$ is generated as a ring extension of A by the images of x and y subject to the relations

$$ya = \alpha(a)y; \quad xa = \beta(a)x; \quad xy = u; \quad yx = \alpha(u).$$

(We abuse notation and omit - in the factor.) We shall denote this ring, which is independent of the parameter ρ, by T or, more explicitly, by $T(A, \alpha, u)$. Localizing T at the powers of y gives the skew Laurent polynomial ring $A[y^{\pm 1}; \alpha]$ which is itself a special case of T with $u = 1$.

The rings R arising in our general construction in 1.1 can be viewed as rings of this form T, namely $T(R_0, \alpha, w)$ with $R_0 = A[w;\gamma]$ and $\alpha(w) = \rho^{-1}w - v$ or, in the conformal case, $R_0 = A[z;\gamma]$ and $\alpha(z) = \rho^{-1}z$, as in 1.8. This observation gives a method for generalizing results from the conformal case with $\rho = 1$ by interpretting them with A replaced by R_0 and passing to the factor. For an example of this, see [16, Theorem 3.12] where a result on the semisimplicity of finite-dimensional R-modules is extended from the case $\rho = 1$ so as to apply to such examples as those described in 1.2(iii) and (iv).

Rings of this form T, with u central, have been independently studied by Bavula in a sequence of papers including [2, 3, 4]. For the case where A is commutative, necessary and sufficient conditions for T to be primitive are given in [12].

2. Simple modules

In this section we discuss some simple left R-modules, beginning with some which generalize the Verma modules over $U(sl_2)$. Let M be a maximal left ideal of A and set $V(M) := \dfrac{R}{RM + Rx}$. In the theorem below the v_n are as specified in 1.7. In particular, in the conformal case, $v_n = (u - \rho^n\alpha^n(u))$.

2.1. Theorem. *The module $V(M)$ is simple if and only if $u - \rho^n\alpha^n(u) \notin \alpha^{n-1}(M)$ $\forall n > 0$.*

Proof. Let $I = RM + Rx$. Every element of R/I can be written in the form $(\sum_{i\geq 0} y^i a_i) + I$ where the $a_i \in A$ are unique modulo M. The action of x on R/I is given by 1.7(iii): if $n > 0$,

$$x(y^n a + I) = v_n y^{n-1}a + I = y^{n-1}\gamma(a)\alpha^{1-n}(v_n) + I.$$

If $v_n \in \alpha^{n-1}(M)$ for some n then $x(y^n A + I) = 0$, and hence there is a non-zero proper submodule $\sum_{j\geq n}(y^n A + I) = (Ry^n + I)/I$ so that R/I is not simple. Conversely, suppose that $v_n \notin \alpha^{n-1}(M)\ \forall n > 0$ and let X be a non-zero submodule of R/I. Choose a non-zero element $t \in X$ of the form $(y^n a_n + \ldots + a_0) + I$ with $a_n \notin M$ and n minimal. Suppose that $n > 0$. Then $xt = \sum_{0<i\leq n} y^{i-1}\gamma(a_i)\alpha^{1-i}(v_i) + I$. By the choice of n, $\gamma(a_n)\alpha^{1-n}(v_n) \in M$. But $\{a \in A : \gamma(a)\alpha^{1-n}(v_n) \in M\}$ is a left ideal of A containing M and, as $v_n \notin \alpha^{n-1}(M)$, it is proper and equal to M. Hence $a_n \in M$, a contradiction unless $n = 0$. Thus $n = 0$ and $a_0 + I \in X$ where $a_0 \notin M$. It follows that $1 + I \in X$ whence $X = R/I$ and R/I is simple. □

2.2. Theorem. *If $V(M)$ is not simple and $d > 0$ is minimal with $v_d \in \alpha^{d-1}(M)$ then $L(M) := R/(RM + Rx + Ry^d)$ is simple.*

Proof. This is proved by adapting the proof of 2.1 with I now the left ideal $RM + Rx + Ry^d$. □

2.3. Examples. (i) We have seen in 1.7 that if q is not a root of unity then $k[y]$, which was identified with $\frac{A_1^q}{A_1^q x}$, is a simple A_1^q-module. Its simplicity is an example of 2.1 with $A = k$, $M = 0$ and $v_n = \frac{q^n - 1}{q - 1}$. Of course, the same conclusion holds when $q = 1$, where $v_n = n$. In the construction of $R = A_2^{\overline{q},\lambda}$, where $A = A_1^{q_1}$ and $v_n = \frac{q_2^n - 1}{q_2 - 1} z_1$, taking $M = x_1 A_1^q$ yields, provided q_1 and q_2 are not roots of unity, a simple R-module, $V(M) = \frac{R}{Rx_1 + Rx_2}$ which can be identified with $k[y_1, y_2]$. The actions of the four generators on this module are given in 1.7. Repeated application of 2.1 in this way gives natural simple $A_n^{\overline{q},\Lambda}$-modules under appropriate conditions on the parameters.

(ii) Probably the best examples to illustrate the case where $V(M)$ is not simple come from the Verma modules over $U(sl_2)$ which are the prototype for the modules $V(M)$. Here $A = k[t]$ and $v_n = u - \rho^n\alpha^n(u) = n(t+n-1)$. Suppose that char$k = 0$. There is a unique maximal ideal M_n, generated by $t - n + 1$, with $v_n \in \alpha^{n-1}(M_n)$. For these maximal ideals M_n, M_n has an n-dimensional simple factor $L(M_n)$ but for other maximal ideals M, $V(M)$ is simple.

For the dispin algebra, 1.2(iii), v_n is $-n/2$ if n is even and $t + \frac{(n-1)}{2}$ if n is odd. This gives rise to one n-dimensional simple R-module for each odd positive integer n.

For the Woronowicz example, 1.2(iv), v_n is again linear in t and, provided ν is not a root of unity, there is a countable set $\{M_n\}$ of maximal ideals for which $V(M_n)$ has an n-dimensional simple factor $L(M_n)$. This example differs from the previous two in that there are other finite-dimensional simple modules. These are all one-dimensional and arise from the existence of a maximal ideal $M = tA$ such that $\alpha(M) = M$. We now proceed to discuss a class of modules of which they are examples.

2.4. Notation. From now on we shall work with the conformal case and make use of the normal element z. The same methods can be applied in general with z replaced by w but the simplicity criteria are more complex. With M still denoting a maximal left ideal of A and with $R_0 = A[z : \gamma]$ as in 1.8, we consider R-modules which are induced from a simple R_0-module of the form $\dfrac{R_0}{R_0M + R_0(z-\eta)}$ for some $\eta \in k$. If $\eta \neq 0$ and $M \neq \gamma(M)$ then there exist $m, n \in M$ with $1 = m + \gamma(n)$ so that $\eta = zn + (m-1)(z-\eta) + \eta m \in R_0M + R_0(z-\eta)$ and hence $R_0M + R_0(z-\eta) = R_0$. On the other hand, if $\gamma(M) = M$ and $\eta \neq 0$ then, as can be seen from Proposition 2.5 below or by direct calculation, $R_0M + R_0(z-\eta)$ is a maximal left ideal of R_0. This leads us to consider the case where $\eta \neq 0$ and $\gamma(M) = M$, when we set $W(M,\eta) := \dfrac{R}{RM + R(z-\eta)}$. In determining when $W(M,\eta)$ is simple, we first decide when its localization at the powers of y is a simple left R_y-module. As $R_y = R_0[y, y^{-1}; \alpha]$, the following result is applicable.

2.5. Proposition. *Let S be a ring with an automorphism α and a maximal left ideal M and let $T = S[y, y^{-1}; \alpha]$.*

(i) *The left T-module $\dfrac{T}{TM}$ is simple if and only if for all positive integers n and all $a \in S \backslash M$, $\alpha^n(M)a \not\subseteq M$.*

(ii) *If $\dfrac{T}{TM}$ is not simple and n is a positive integer minimal such that there exists $a \in S\backslash M$ with $\alpha^n(M)a \subseteq M$ then the left ideal $J = TM + T(y^n - a)$ is maximal.*

Proof. (i) See [8, Lemmas 3.4 and 3.5] (where right modules are considered).

(ii) Let $I = \{r \in S : ra \in M\}$, a left ideal of S containing $\alpha^n(M)$. As $\alpha^n(M)$ is maximal and $a \notin M$, $I = \alpha^n(M)$.

Suppose that $TM + T(y^n - a) = T$ and choose $t = \sum_{i=k}^{m} y^i r_i$ in T with minimal length $m - k + 1$ such that $t(y^n - a) - 1 \in TM$. If $m > -n$ then, by inspection of terms of degree $m + n$, $\alpha^{-n}(r_m) \in M$ and, as $I = \alpha^n(M)$, $r_m a \in M$, whence $y^m r_m(y^n - a) \in TM$. Similarly, if $k < 0$ then $r_k a \in M$, $\alpha^{-n}(r_k) \in M$ and $y^k r_k(y^n - a) \in MT$. Each of these possibilities contradicts the minimality of t so $-n \geq m \geq k \geq 0$ which is impossible. Thus J is proper.

As $y^n - a \in J$ and $y^{-1} - y^{n-1}\alpha^{-n}(b) \in J$, where $b \in S$ is such that $1 + ba \in M$, every element of T/J can be written in the form $(\sum_{0\leq j\leq n-1} y^j s_j) + J$, $s_j \in S$. Replacing 1 in the above argument by an element of $S[y;\alpha]$ of degree at most $n-1$ shows that the s_j are unique modulo M. If X is a non-zero submodule then we can choose $h = (\sum_{0\leq j\leq n-1} y^j s_j) + J \in X$ with a minimal number of non-zero terms. By the action of y and the simplicity of S/M, we can suppose that $s_{n-1} = 1$. Suppose that there is another non-zero term, $y^i s_i + J$, say, in h. Thus $s_i \notin M$. For $m \in M$, the term $y^{n-1}m + J$ in $\alpha^{n-1}(m)h$ is 0. By minimality, $\alpha^{n-1}(m)h = 0$ and so $y^i\alpha^{-i+n-1}(m)s_i + J = 0$, whence $\alpha^{-i+n-1}(M)s_i \subseteq M$, contradicting the minimality of n. Thus $y^{n-1} + J \in X$ and it follows that $X = T/J$ which is therefore simple. □

2.6. Theorem. *Let M be a maximal left ideal of A such that $\gamma(M) = M$ and let $0 \neq \eta \in k$. The left R-module $W(M, \eta)$ is simple if and only if the following conditions hold:*

(a) $\rho^j\eta + u \notin \alpha^j(M)\ \forall j$;

(b) *there do not exist* $a \in A\backslash M$ *and* $n > 0$ *with both* $\alpha^n(M)a \subseteq M$ *and* $\gamma(a) - \rho^n a \in M$.

2.7. Remark. Before proving 2.6, we make some remarks on Condition (b). If $N = R_0M + R_0(z - \eta)$ and there exists a positive integer n with $\alpha^n(N)a \subseteq N$ for some $a \in R_0\backslash N$ then a can be assumed to be in A. Condition (b) interprets the condition $\alpha^n(N)a \subseteq N$ in terms of A rather than R_0, given that $\eta \neq 0$. It is equivalent to the non-existence of any R_0-isomorphisms between R_0/N and $R_0/\alpha^n(N)$ with $n \neq 0$. In the case where A is commutative and $\gamma = \mathrm{id}_A$, it becomes much simpler, namely there does not exist $n > 0$ with both $\alpha^n(M) = M$ and $\rho^n = 1$.

Proof of 2.6. Let $I = RN = RM + R(z - \eta)$. Thus $R/I = W(M, \eta)$, which we now abbreviate to W. Using the formulae 1.8(i) and (ii), every element of W can be written in the form $(\sum_{i\geq 0} y^i a_i + \sum_{i<0} x^{-i}a_i) + J$ where the $a_i \in A$ are unique modulo M. For $i \geq 0$, $yx^ia = (\rho^{-1}z + \alpha(u))x^{i-1}a = x^{i-1}(\rho^{-i}z + \alpha^i(u))a = x^{i-1}\gamma(a)(\rho^{-i}z + \alpha^i(u))$, we have

$$y(x^ia + J) = x^{i-1}\gamma(a)(\rho^{-i}\eta + \alpha^i(u)) + J.$$

Similarly, for $i \geq 0$,

$$x(y^ia + J) = y^{i-1}\gamma(a)(\rho^{i-1}\eta + \alpha^{1-i}(u)) + J.$$

If (b) fails then, by 2.5 and Remark 2.7 above, the localization W_y is not simple, whence W is not simple. Suppose that (a) fails, that is $\rho^j\eta + u \in \alpha^j(M)$ for some integer j. If $j \geq 0$ then $xy^{j+1} + J = 0$ and $\sum_{i>j}(y^iA + J)$ is a proper non-zero submodule, with factor $\dfrac{R}{RN + Ry^{j+1}}$. If $j < 0$ then $yx^{-j} = 0$ and $\sum_{i\leq j}(x^{-i}A + J)$ is a proper non-zero submodule, with factor $\dfrac{R}{RN + Rx^{-j}}$. Thus W is not simple.

Suppose now that both (a) and (b) hold. By (b) and 2.5, the localization W_y is simple. So if $X = J/I$ is a non-zero submodule of $W = R/I$ then either X is torsion with respect to $\{y^i\}$ or $X_y = W_y$. From the action of y given above, the former can only occur if there exist $i > 0$ and $a \in A\backslash M$ such that $yx^ia \in J$, that is, $\gamma(a)(\rho^{-i}\eta + \alpha^i(u)) \in M$. Let $H = \{r \in A : r(\rho^{-i}\eta + \alpha^i(u)) \in M\}$, a left ideal of A. The condition $\gamma(M) = M$ ensures both that $M \subseteq H$ and that $\gamma(a) \notin M$, whence $H = A$. Thus $\rho^{-i}\eta + \alpha^i(u) \in M$, contradicting (a). If $X_y = W_y$ and $X \neq W$ then $y^j \in J$ for some $j \geq 1$. For the minimal such value of j we have

$$y^{j-1}(\rho^{1-j}z - \alpha^{1-j}(u)) = (z - u)y^{j-1} = xy^j,$$

whence $y^{j-1}(\rho^{1-j}\eta - \alpha^{1-j}(u)) \in J$. By the simplicity of the A-module A/M and the minimality of j, $(\rho^{1-j}\eta - \alpha^{1-j}(u)) \in M$, again contradicting (a). Thus W is simple. □

2.8. Theorem. *If condition* (b) *in Theorem* 2.6 *fails and n is the least positive integer for which there exists $a \in A\backslash M$ with $\alpha^n(M) \subseteq M$ and $\gamma(a) - \rho^n a \in M$ then $B(M,\eta,a) := R/(R(y^n - a) + RM + R(z-\eta))$ is simple.*

Proof. Let $B = B(M,\eta,a)$. With $N = R_0M + R_0(z-\eta)$ as before, n is the least value of n for which there exists $a \in R_0\backslash N$ with $\alpha^n(N)a \subseteq N$. It follows from 2.5 that the localization B_y is simple. Let $J = RN + R(y^n - a)$, so that $B = R/J$. Let $X = I/J$ be a submodule of B. Then X is y-torsion or $y^i \in I$ for some i. Every element of B can be written in the form $(\sum_{0\leq j\leq n-1} y^j a_j) + J$ with the a_j being unique modulo M. For $b \in A$, $y(y^{n-1}b + J) = \alpha^n(b)a + J$. By the choice of a, $\alpha^n(b)a \in M$ if and only if $b \in M$. It follows that if X is y-torsion then $X = 0$. If $y^i \in I$ then the above rule for the action of y on $y^{n-1}b + J$ shows that $y^i \equiv y^r b_r \bmod J$ where $i \equiv r \bmod n$, $0 \leq r < n$ and $b_r \notin M$. As $Ab_r + M = A$ and $M \subseteq I$, it follows that $y^r \in I$ and hence that $y^n \in I$ and $a \in I$. But $M \subseteq I$ and $M + Aa = A$ so $I = R$. Thus B is simple. □

2.9. Remark. In 2.4, 2.6 and 2.8 we have assumed that the parameter η is non-zero. The results extend to the case $\eta = 0$ where it is no longer necessary to assume that $\gamma(M) = M$. For all maximal left ideals M of A, the left ideal $N = R_0M + R_0z$ is maximal. The following theorem, whose proof is easily adapted from those of 2.6 and 2.8, gives the condition for R_0/N to induce a simple left R-module.

2.10. Theorem. *Let M be a maximal left ideal of A.*

(i) *The left R-module $W(M,0) = \dfrac{R}{RM + Rz}$ is simple if and only if the following conditions hold:*

(a) $u \notin \alpha^j(M)\ \forall j$.

(b) *there do not exist $a \in A\backslash M$ and $n > 0$ with $\alpha^n(M)a \subseteq M$.*

(ii) *If condition* (b) *in* (i) *fails and n is the least positive integer for which there exists $a \in A\backslash M$ with $\alpha^n(M) \subseteq M$ then $B(M,0,a) := R/(R(y^n - a) + RM + Rz)$ is simple.*

2.11. Remark. When the module $W(M,\eta)$ is not simple there are simple factors, analogous to those given in 2.8 and 2.10(ii), but with x replacing y. When $\gamma(M) = M$ and $N = R_0M + R_0(z-\eta)$, $\beta^n(N) = \alpha^{-n}(N)$ and, for a positive integer n, there exists $a \in A\backslash M$ such that $\alpha^n(N)a \subseteq N$ if and only if there exists $b \in A\backslash M$ such that $\alpha^{-n}(N)b \subseteq N$. In the situation of 2.8 there are simple modules of the form $B^*(M,\eta,b) := \dfrac{R}{(R(x^n - b) + RM + R(z-\eta))}$. In the case $\eta = 0$ with $\gamma(M)$ not necessarily equal to M, there is a simple module $B^*(M,0,a) := R/(R(x^n - b) + RM + Rz)$ if n is minimal such that $\beta^n(M)b \subseteq M$ for some $b \in A\backslash M$.

2.12. Examples. (i) When A is commutative γ is necessarily the identity so the condition $\gamma(M) = M$ is redundant. For all maximal ideals M and all $\eta \in k$ we can consider $W(M,\eta) = \dfrac{R}{RM + R(z-\eta)}$. Combining the commutative version of 2.6(b) with 2.10(b) for the case $\eta = 0$, $W(M,\eta)$ is simple if and only if 2.6(a) holds and there does not exist $n > 0$ with both $\alpha^n(M) = M$ and $\eta(\rho^n - 1) = 0$. For $U(sl_2)$ in characteristic 0, where $\rho = 1$, $\alpha^n(M) \neq M$ for all M and all $n > 0$. For each j and $\eta \neq 0$ (resp. $\eta = 0$), there are two (resp. one) maximal ideals M with $\eta + u \in \alpha^j(M)$. Apart from these cases, $W(M,\eta)$ is simple. On the other hand, in finite characteristic $W(M,\eta)$ is never simple. The dispin algebra, 1.2(iii), is similar except that each η and j determine one maximal ideal M where (a) fails. For the Woronowicz example, 1.2(iv), there is one maximal ideal, $M = tA$, with $\alpha(M) = M$, giving rise to simple factors $B(M, 0, \lambda)$ for all $\lambda \in k^*$. If ν is a root of unity there are simple factors $B(M,\eta,\lambda)$ for all maximal ideals M and all η.

When A is commutative, as in these examples, and k is algebraically closed, one can classify all the finite-dimensional simple R-modules. Every finite-dimensional simple left R-module is isomorphic to one of the modules $L(M)$, $B(M,\eta,a)$ or $B^*(M,\eta,a)$. See [14] for details.

(ii) Consider the case of A_1^q where M is necessarily 0. Suppose that q is not a root of unity. Then 2.6(b) holds for all $\eta \neq 0$ and 2.6(a) holds if and only if $q^j\eta + \frac{1}{1-q} \neq 0$ for all j. Consequently $W(0,\eta) = \dfrac{A_1^q}{A_1^q(z-\eta)}$ is simple for all but countably many values of η, namely $\eta = 0$ and $\eta = \dfrac{1}{q^j(q-1)}$, where j is an integer. For $\eta = \dfrac{1}{q^j(q-1)}$, if $j \geq 0$ then $\dfrac{A_1^q}{A_1^q(z-\eta) + A_1^q y^{j+1}}$ is simple and if $j < 0$ then $\dfrac{A_1^q}{A_1^q(z-\eta) + A_1^q x^{-j}}$ is simple.

(iii) Here we intend to give some genuinely noncommutative examples of the failure of 2.6(b). To be more precise, we give examples where this condition fails but M is not periodic under the action of α. To obtain such examples we consider the rings $A = k[y][x;\sigma]$ and $S = k[y^{\pm 1}][x^{\pm 1};\sigma]$ where $\sigma(y) = qy$. Thus $xy = qyx$, A is the coordinate ring of the quantum plane and S is the two-variable case of the algebra considered by McConnell and Pettit in [18]. The next result gives a supply of maximal left ideals of A and S.

2.13. Proposition. *Let A and S be as above and suppose that $q \in k^*$ is not a root of unity.*

(i) *The following are maximal left ideals of S:*

$$S(1+x); \quad S(1+y); \quad S(1+x^iy^j),$$

where i and j are non-zero coprime integers.

(ii) *The following are maximal left ideals of A:*

$$A(1+x); \quad A(1+y); \quad A(1+x^iy^j),$$

where i and j are positive coprime integers.

Proof. (i) The maximality of the first two follows easily from 2.5. For the third, there is a k-automorphism of S given by $x \mapsto x^i y^j$ and $y \mapsto x^{-s} y^t$ where s and t are integers such that $si + tj = 1$.

(ii) This follows routinely from (i). □

We consider now a ring R constructed from the conformal case of 1.1 with the following data: A is as in 2.13 but with x, y and q renamed as x_1, y_1 and q_1; $\alpha(x_1) = q_2 x_1$, $\alpha(y_1) = q_3 y_1$; u is any non-zero scalar multiple of $x_1 y_1$ and $\rho = q_4$, where the q_i are non-zero scalars. The rings R formed in this way were discussed in 1.6 and include the various parametrizations of $\mathcal{O}_q(M_2(k))$. Let $M = A(1 + x^i y^j)$ where either i and j are coprime positive integers, $i = 1$ and $j = 0$ or $i = 0$ and $j = 1$. By 2.13(ii), M is a maximal left ideal of A. Finally, let $a = x_1^m y_1^r$ where m and r are positive integers. Clearly $a \notin M$.

Consider condition 2.6(b) in this situation. Here $\gamma(x_1) = q_1 x_1$ and $\gamma(y_1) = q_1^{-1} y_1$ so $\gamma(a) - \rho^n a = (q_1^{m-r} - q_4^n)a \in M$ if and only if $q_1^{m-r} - q_4^n = 0$. Also

$$\alpha^n(M)a = (1 + q_2^{in} q_3^{jn} x_1^i y_1^j)aA = a(1 + q_1^{-mj+ri} q_2^{in} q_3^{jn} x_1^i y_1^j)A.$$

Thus $\alpha^n(M)a \subseteq M$ if and only if $q_1^{-mj+ri} q_2^{in} q_3^{jn} = 1$. Any choice of the parameters for which $q_1^{-mj+ri} q_2^{in} q_3^{jn} = 1$ and $(q_1^{m-r} - q_4^n) = 0$ for some n, gives examples where 2.6(b) fails and there are simple factors $B(M, \eta, a)$ of the modules $W(M, \eta)$. For an easy example, take $i = j = m = n = 1, r = 0$ that is $q_4 = q_1 = q_2 q_3$.

2.14. Remark. The above methods apply also to skew Laurent polynomial rings and relate to the discussion of GK-dimension of simple modules in [18]. Let $S = k[y^{\pm 1}][x^{\pm 1}; \sigma]$ where $\sigma(y) = q_1 y$, let α be the k-automorphism of S such that $\alpha(x) = q_2 x$, $\alpha(y) = q_3 y$ and let $T = S[z^{\pm 1}; \alpha]$, the three variable case of the k-algebras considered in [18]. Let G be the subgroup of k^* generated by the parameters q_1, q_2 and q_3. If G has maximal rank 3 then, by [18, 3.10 and 6.2], every simple T-module has GK-dimension 2. Here we indicate a proof that the converse is true.

Suppose that q_1 is not a root of unity but that there exist non-zero integers r, s, t such that $q_1^r = q_2^s q_3^t$. If m is the highest common factor of s and t, $s = im$ and $t = jm$ then i and j are coprime. Hence $M = S(1 + x^i y^j)$ is a maximal left ideal of S, by 2.13(i), and $1 = ci + dj$ for some integers c, d. Then $\alpha^m(M) = S(1 + q_1^r x^i y^j)$ and $(1 + x^i y^j)x^{-rc} y^{rd} = x^{-rc} y^{rd}(1 + q_1^r x^i y^j)$. By 2.5, the left T-module T/TM is not simple and T has a simple module of the form $\dfrac{T}{TM + T(z^n - a)}, a \in S \backslash M$. This simple module has the same GK-dimension as S/M, namely 1. Although the above calculation does not cover all cases where G has rank < 3, other cases can be treated in a similar way to obtain the following result.

2.15. Proposition. *All simple T-modules have GK-dimension 2 if and only if the group G has maximal rank 3.*

For a specific example in which T is simple and has a simple module of GK-dimension 1, take $q_1 = 6, q_2 = 2, q_3 = 3$, so that $r = s = t = m = i = j = 1$ and we can take $c = 0, d = 1$. Then, with $M = S(1 + xy)$ and $a = y$, $\dfrac{T}{TM + T(z - y)}$ is simple of GK-dimension 1. In this example, $\dfrac{T}{T(1 + x)}$ is simple of GK-dimension 2.

2.16. Remark. Proposition 2.15 does not extend to the case of 4 variables, where the maximal rank of the group generated by the parameters is 6. This can be seen from [18, Example 3.10 and Theorem 6.2].

3. Dimension

Known results on Krull dimension for two well-studied rings, the Weyl algebra A_n and the enveloping algebra $U(sl_2)$ can be extended.

3.1. Theorem. *If no q_i is a root of unity then the localization of $A_n^{\bar{q},\Lambda}$ at the multiplicatively closed set Z generated by the z_i's has Krull and global dimension n.*

The proof of this theorem is given in [15]. Here we sketch the proof for the case $n = 1$. Let R be the localization of A_1^q at the powers of $z = xy - yx$. The isomorphic localizations $R_y = k[z^{\pm 1}][y^{\pm 1}; \gamma]$, where $\gamma(z) = q^{-1}z$, and R_x are well-known to have Krull dimension 1 and it therefore suffices to show that $R_y \oplus R_x$ is a faithfully flat extension of R. To do this let M be a non-zero left R-module which is both y-torsion and x-torsion. There exists a minimal positive integer n for which $xm = y^n m = 0$ for some $0 \neq m \in M$. Formula 1.7(v) then gives $y^{n-1}m = 0$, contradicting the choice of n.

Smith [20] showed that the Krull dimension of $U(sl_2)$ is 2 when $k = \mathbb{C}$. This is generalized in [13] where the following is proved in the conformal case with $\rho = 1$. This improved on an earlier proof in [10] which required additional hypotheses.

3.2. Theorem. *Suppose that A is α-simple and a principal ideal domain. Then the Krull dimension of R is 2.*

The method again involves $R_y \oplus R_x$ but here it need not be faithfully flat, there often being finite-dimensional simple modules which are both y-torsion and x-torsion. However if M is a finitely generated left R-module such that both M_y and M_x have finite length then M itself has finite length and this suffices to prove the theorem.

References

[1] M. Artin, W. Schelter and J. Tate, Quantum deformations of GL_n, Communications on Pure and Applied Mathematics 44 (1991), 879–895.

[2] V. V. Bavula, Generalized Weyl algebras and their representations, St Petersburg Math. J. 4 (1993), 71–92.

[3] V. V. Bavula, Description of bilateral ideals in a class of noncommutative rings I, Ukrainskii Matematicheskii Zhurnal 45 (1993), 209–220.

[4] V. V. Bavula, Generalized Weyl algebras, kernel and tensor-simple algebras, their simple modules, Canadian Math. Soc. Conference Proceedings 14 (1993), 83–107.

[5] R. Dipper and S. Donkin, Quantum GL_n, Proc. London Math. Soc. (3) 63 (1991), 165-211.

[6] L. D. Faddeev, N. Yu. Reshetikhin and L. A. Takhtadzhyan, Quantization of Lie groups and Lie Algebras, Leningrad Math. J. 1 (1990), 193–225.

[7] K. R. Goodearl, Prime ideals in skew polynomial rings and quantized Weyl algebras, J. Algebra 150 (1992), 324–377.

[8] T. Hodges and J. C. McConnell, On Ore and skew-Laurent extensions of Noetherian rings, J. Algebra 73 (1981), 56–64.

[9] D. A. Jordan, Iterated skew polynomial rings and quantum groups, J. Algebra 156 (1993), 194–218.

[10] D. A. Jordan, Krull and global dimension of certain iterated skew polynomial rings, Abelian Groups and Noncommutative Rings, a Collection of Papers in Memory of Robert B. Warfield, Jr., Contemporary Mathematics 130 , Amer. Math. Soc. (1992), 201–213.

[11] D. A. Jordan, Height one prime ideals of certain iterated skew polynomial rings, Math. Proc. Camb. Philos. Soc. 114 (1993), 407–426.

[12] D. A. Jordan, Primitivity in skew Laurent polynomial rings and related rings, Math. Z. 213 (1993), 353–371.

[13] D. A. Jordan, Iterated skew polynomial rings of Krull dimension two, Archiv der Mathematik 61 (1993), 344–347.

[14] D. A. Jordan, Finite dimensional simple modules over certain iterated skew polynomial rings, J. Pure Appl. Algebra 98 (1995), 45–55.

[15] D. A. Jordan, A simple localization of the quantized Weyl algebra, J. Algebra 174 (1995), 267–281.

[16] D. A. Jordan and I. E. Wells, Invariants for automorphisms of certain iterated skew polynomial rings, preprint, University of Sheffield.

[17] V. G. Kac, Lie superalgebras, Adv. Math. 26 (1977), 8-96.

[18] J. C. McConnell and J. J. Pettit, Crossed products and multiplicative analogues of Weyl algebras, J. London Math. Soc. (2) 38, 1988, 47–55.

[19] S-Q. Oh, Primitive ideals of the coordinate ring of quantum symplectic space, J. Algebra 174 (1995), 531–552.

[20] S. P. Smith, Krull dimension of the enveloping algebra of $sl(2, \mathbb{C})$, J. Algebra 71 (1981), 189–194.

[21] S. P. Smith, Quantum groups. An introduction and survey for ring theorists, Noncommutative rings, MSRI Pub. 24 (1992), Springer, Berlin, Heidelberg, New York.

[22] M. Vancliff, Quadratic algebras associated with the union of a quadric and a line in $\mathbb{P}^3$, preprint, University of Washington.

[23] I. E. Wells, Simplicity in some iterated skew polynomial rings, preprint, University of Sheffield.

[24] S. L. Woronowicz, Twisted $SU(2)$-group. An example of a non-commutative differential calculus, Publ. R.M.I.S., Kyoto Univ. 23 (1987), 117–181.

The completeness principle for the Golay codes and some related codes

Helmut Koch

The words of weight 8 of the binary Golay code G of length 24 form a $S(5, 8, 24)$ Steiner system. One generalization of this well known fact to higher weights is the theorem that the set of words of G of given weight form a 5-design, which is a special case of the Theorem of Assmus and Mattson [1]. We consider here an other generalization which we call the "completeness principle" and we study this principle also for other codes.

For any code $C \subset \mathbb{F}_q^n$ we denote by C_m the set of words in C of weight m. In particular for binary codes we use the set-theoretical notation, i.e. we identify an element of $\mathbb{F}_2^n$ with its support.

1.

It is well known that the set system G_8 for the binary Golay code G of length 24 is a Steiner system, i.e. given any set a of 5 positions of G, there is exactly one word $x \in G_8$ with $a \subset x$.

To prove this property it is sufficient to know that the Golay code is a linear code with weight enumerator

$$1 + 759X^8 + 2576X^{12} + 759X^{16} + X^{24}.$$

In fact there is at most one $x \in G_8$ with $a \subset x$. On the other hand every $x \in G_8$ contains $\binom{8}{5}$ sets of places of cardinality 5. Since one has alltogether

$$\binom{24}{5} = 759\binom{8}{5}$$

such sets, it follows that there is an $x \in G_8$ with $a \subset x$ for every set a of 5 places of G.

Now we ask whether there is something similar for the sets G_{12} and G_{16}. If a is a set of 9 places of G. It is clear that there is at most one $x \in G_{12}$ with $a \subset x$. On the other hand not for all such a there can exist some $x \in G_{12}$ with $a \subset x$. Namely, if $|a \cap y| \in \{7, 8\}$ for a word $y \in G_8$ (considered as subset of the set of

places of G), then $a \subset x$ for $x \in G_{12}$ would imply $|x \cap y| = 8$ which is impossible. The number of sets a of th
is kind is

$$|G_8|(16 + 8\binom{16}{2}) = 740784.$$

The number of sets a with $a \subset x$ for $x \in G_{12}$ is

$$|G_{12}|\binom{12}{9} = 566720.$$

The sum of these numbers is $\binom{24}{9}$. This proves the following "completeness principle" for G_{12}:

Proposition 1.1. *For any set a of 9 places of the Golay code G with the property $|a \cap y| < 7$ for every $y \in G_8$ there exists one and only one $x \in G_{12}$ with $a \subset x$.*

The corresponding completeness principle for G_{16} is as follows:

Proposition 1.2. *For any set a of 13 places of the binary Golay code G with the property*

$$|a \cap y| < 11 \, \textit{for every} \quad y \in G_{12}$$

there exists one and only one $x \in G_{16}$ with $a \subset x$.

2.

Now we consider the ternary Golay code $T \subset \mathbb{F}_3^{12}$. Again the completeness principle for minimal weight is well known:

Given a set a of 5 places of T there is one and only one pair $x, -x$ of words in T_6 such that the components x_i of x for $i \in a$ are not equal to 0.

T has 440 words of weight 9. Given any set a of 9 places, it is clear that there is not more than one pair $\{x, -x\}$ of words in T_9 with support in a. Since $\binom{12}{9} = 220$ there is exactly one pair $\{x, -x\}$ of words in T_9 with support in a.

3.

We come back to binary codes and denote by $\mathfrak{C}_{24}$ the set of all linear doubly-even self dual codes in $\mathbb{F}_2^{24}$. It is well known that beside the Golay codes there are up to equivalence exactly 8 such codes C. They are characterized by their tetrad systems C_4.

Following [3] we denote by d_{2n} the linear code in $\mathbb{F}_2^{2n}$ generated by $\{1,2,3,4\}$, $\{1,2,5,6\}, \ldots, \{1,2,2n-1,2n\}$. Furthermore, $e_7 \subset \mathbb{F}_2^7$ is the linear code generated by d_6 and $\{1,3,5,7\}$ and $e_8 \subset \mathbb{F}_2^8$ is the linear code generated by d_8 and $\{1,3,5,7\}$. The codes C in $\mathfrak{C}_{24}$ with $C_4 \neq \phi$ have the tetrad systems $6d_4$, $4d_6$, $3d_8$, $2d_{12}$, d_{24}, $d_{16}+e_8$, $3e_8$, $d_{10}+2e_7$.

Each of these codes satisfies completeness principles. We restrict in this section to weight 8.

Let $C \in \mathfrak{C}_{24}$ be a code with tetrad system $6d_4$. A word in C has the form $(c_1,\ldots,c_8)$, where $c_i \in d_4^\perp$. With $(c_1,\ldots,c_8$ to C belong the words $(c_1+x_1,\ldots,c_6+x_6)$, where $x_i \in d_4$.

Since the group $d_4^\perp/d_4$ is isomorphic to $\mathbb{F}_4^+$ we introduce the notation

$$0 := \phi,\ \ 1 := \{1,2\},\ \ \rho := \{1,3\},\ \ \rho^2 := \{1,4\},$$

where ρ is a generator of $\mathbb{F}_4^\times$.

Any word of weight 8 in C, which is not in $6d_4$, has up to permutation of its components the form

$$(*) \qquad a = (a_1,a_2,a_3,a_4,0,0) \ \text{ with } \ a_i \in d_4^\perp,\ |a_i| = 2.$$

Up to equivalence we can assume that a has the form $(1,1,1,1,0,0)$. If C contains a further element b of the form $b = (b_1,b_2,b_3,b_4,0,0)$ with $b_i \in d_4$, $|b_i| = 2$, then $b_i \not\equiv 1 (\text{mod } d_4)$ since otherwise $a+b$ would have a weight smaller than 8. Hence up to equivalence b has the form $(\rho,\rho,\rho,\rho,0,0)$ and we have the further word $a+b = (\rho^2,\rho^2,\rho^2,\rho^2,0,0)$ in C. It is clear that up to congruence mod d_4 there cannot be more words of type $(*)$ in C. Alltogether this gives $3\cdot 2^4$ words of type $(*)$ in C and since we have 15 possibilities for the distribution of zeros these are $15\cdot 3\cdot 2^4$ words. On the other hand $6d_4$ contains 15 words of weight 8, which shows that C cannot have more than 735 words of weight 8. But since $|C_8| = 735$ we have the following completeness principle.

Proposition 3.1. *Let C be a code in $\mathfrak{C}_{24}$ with tetrad system $6d_4$ and let a be an arbitrary subset of $\{1,2,3,4,5,6\}$ of cardinality 4. Then there are exactly 3 classes $c = (c_1,\ldots,c_6)$ of minimal weight 8 in $C/6d_4 \subset 6d_4^\perp/d_4$ with non zero components $\bar{c}_i$ for $i \in a$.*

For the proof of Proposition 3 we needed only the fact that there are 735 words in C_8 which follows from Gleasons's Theorem [4]. Instead of this theorem we can also use the balance principle [5]: Let $C \subset \mathbb{F}_2^n$ be a self-dual linear code, let a be a set of places and $b = 1+a$, $C_a = \{x \in C | x \subseteq a\}$ Then

$$\dim C_a - \frac{|a|}{2} = \dim C_b - \frac{|b|}{2}.$$

In our case let a be the support of $2d_4$, $C_a = 2d_4$. Then $\dim C_a = 2$ and therefore $\dim C_b = 6$, $\dim C_b/4d_4 = 2$.

Since an element in C_4 has the form $(0,0,a_1,a_2,a_3,a_4)$ with $a_i \in d_4^\perp$ Proposition 3 follows. In this way one sees that $|C_8| = 735$ without using Gleason's Theorem. One gets similar results in all other cases of codes in $\mathfrak{C}_{24}$. We leave this to the reader.

4.

The completeness principle can be used to prove the uniqueness of the code in question as a doubly-even self-dual linear code with given tetrad system. In the case of the Golay code this may be well known, but I found no reference in the literature.

One proceeds as follows: We choose successively 12 sets $a_1, \dots, a_{12}$ in $\{1, \dots, 24\}$ with cardinality 5 and take a word c_i of weight 8 containing a_i which is orthogonal to all words already chosen, $i = 1, \dots, 12$. The choice of c_i is always unique up to equivalence:

$$\begin{aligned}
a_1 &= \{1,2,3,4,5\}, & c_1 &= \{1,2,3,4,5,6,7,8\},\\
a_2 &= \{1,2,3,4,9\}, & c_2 &= \{1,2,3,4,9,10,11,12\},\\
a_3 &= \{1,2,3,4,13\}, & c_3 &= \{1,2,3,4,13,14,15,16\},\\
a_4 &= \{1,2,3,4,17\}, & c_4 &= \{1,2,3,4,17,18,19,20\},\\
a_5 &= \{1,2,3,4,21\}, & c_5 &= \{1,2,3,4,21,22,23,24\},\\
a_6 &= \{1,2,3,,5,9\}, & c_6 &= \{1,2,3,5,9,13,17,21\},\\
a_7 &= \{1,2,3,6,9\}, & c_7 &= \{1,2,3,6,9,14,18,22\},\\
a_8 &= \{1,2,3,7,9\}, & c_8 &= \{1,2,3,7,9,15,19,23\},
\end{aligned}$$

it follows

$$c_1 + c_3 + c_4 + c_5 + c_6 + c_7 + c_8 = \{1,2,3,8,9,16,20,24\},$$

$$\begin{aligned}
a_9 &= \{1,2,3,5,10\}, & c_9 &= \{1,2,3,5,10,14,19,24\},\\
a_{10} &= \{1,2,3,5,11\}, & c_{10} &= \{1,2,3,5,11,15,20,22\},
\end{aligned}$$

it follows

$$c_1 + c_2 + c_4 + c_5 + c_6 + c_9 + c_{10} = \{1,2,3,5,12,16,18,23\},$$

$$\begin{aligned}
a_{11} &= \{2,3,4,5,9\}, & c_{11} &= \{2,3,4,5,9,14,20,23\},\\
a_{12} &= \{1,2,4,5,9\}, & c_{12} &= \{1,2,4,5,9,15,18,24\}.
\end{aligned}$$

It is easy to see that $c_1, \dots, c_{12}$ are linearly independent. This proves the uniqueness of the Golay code.

Now let C be a code in $\mathfrak{C}_{24}$ with tetrad system $6d_4$. Then the further 6 basis words $c_1, \dots, c_6$ of C can be chosen as follows: First we put zeros at the last both places:

$$\begin{aligned}
c_1 &= (1,1,1,1,0,0)\\
c_2 &= (\rho,\rho,\rho,\rho,0,0)\\
c_1 + c_2 &= (\rho^2,\rho^2,\rho^2,\rho^2,0,0).
\end{aligned}$$

Now we take zeros at the first two places. The components on the third and fourth place have to be equal and it is easy to see that we can choose c_3 and c_4

up to equivalence uniquely as follows:

$$\begin{aligned} c_3 &= (0,0,1,1,1,1) \\ c_4 &= (0,0,\rho,\rho,\rho,\rho) \\ c_3 + c_4 &= (0,0,\rho^2,\rho^2,\rho^2,\rho^2). \end{aligned}$$

Now we take zeros at the first and last place: We can put on the second place an arbitrary element from $\mathbb{F}_4^\times$, so we take 1. Then at the third and fourth place we have to take ρ and ρ^2 in arbitrary order since so far the third and fourth places can be exchanged by equivalence. The component at the fifth place has now to be equal to 1. Hence we can put

$$c_5 = (0,1,\rho,\rho^2,1,0)$$

c_6 of the form $(0,\rho,x_3,x_4,x_5,0)$ is already uniquely determined. One has to put

$$c_6 = (0,\rho,\rho^2,1,\rho,0).$$

We see that our code $C/6d_4$ is even a linear $\mathbb{F}_4$-code. This well known code is called Hexacode, see [3], p. 82.

5.

The completeness principle for binary codes is not restricted to codes in $\mathfrak{C}_{24}$. It is trivially fulfilled for codes in $\mathfrak{C}_8$, $\mathfrak{C}_{16}$. Let us have now a look to $\mathfrak{C}_{32}$. Conway and Pless [2] proved that up to equivalence there are 85 codes in $\mathfrak{C}_{32}$. We consider here as an example the unique code C in $\mathfrak{C}_{32}$ with tetrad system $4d_6 + 2d_4$.

The balance principle shows that some types of elements in $4d_6^\perp/d_6 + 2d_4^\perp/d_4$ cannot be in $C/4d_6 + 2d_4$: Since there is only the zero element in the subspace of $C/4d_6 + 2d_4$ with zero at the first three components, the same is the case for the subspace with the last three components 0. This shows that there is no word of the form

$$(\{1,3,5\},\{1,3,5\},\{1,2\},0,0,0)$$

in C. The subspace of $C/4d_6+2d_4$ with 0 in the last two components has dimension 2 by the balance principle, hence up to equivalence it is generated by the classes of

$$(\{1,3,5\},\{1,3,5\},\{1,3,5\},\{1,3,5\},0,0)$$

and

$$(\{1,2\},\{1,2\},\{1,2\},\{1,2\},0,0).$$

It follows that also the type $(\{1,2\},\{1,2\},\{1,2\},0,\{1,2\},0)$ is impossible. Now consider the subspace of $C/4d_6 + 2d_4$ with 0 in the first two components. The dimension of this space is again 2. Up to equivalence the space is generated by the classes of

$$(0,0,\{1,2\},\{1,2\},\{1,2\},\{1,2\})$$

and

$$(0, 0, \{1, 3, 5\}, \{1, 3, 5\}, \{1, 2\}, 0).$$

We see that in the case $4d_6 + 2d_4$ certain types are forbidden and all other types are realized in $C/4d_6 + 2d_4$.

References

[1] E. F. Assmus, Jr., H. F. Mattson, Jr., New 5-designs, J. Combin. Theory 6 (1969), 122–151.

[2] J. H. Conway, V. Pless, On the enumeration of self-dual codes, J. Combin. Theory A 28 (1980), 26–53.

[3] J. H. Conway, N. J. A. Sloane, Sphere Packings, Lattices and Groups, Grundlehren der mathematischen Wissenschaften 290, Springer-Verlag, New York e.a. 1988.

[4] A. M. Gleason, Weight polynomials of self-dual codes and the MacWilliams identities, Actes Congrès Intern. Math., Gauthier-Villars, Paris 1971, vol. 3, 211–215.

[5] H. Koch, On Self-dual, doubly-even codes of length 32, J. Combin. Theory A 51 (1989), 63–76.

Finite-dimensional Lie algebras with a nonsingular derivation*

A. I. Kostrikin and M. I. Kuznetsov

1. Introduction

The present work is connected with the problems arising in the Lie algebra theory in connection with recent studies on p-groups and pro-p groups of a finite coclass. These investigations were initiated by Leedham-Green and Newman in 1980 [18]. According to their definition a group G of order p^n and nilpotency class c has the coclass $r = n - c$. An inverse limit of finite p-groups of coclass r is called a pro-p group of coclass r. Leedham-Green and Newman made a series of conjectures on the structure of pro-p groups and finite p-groups of a given coclass. These conjectures have gradually been proved by Leedham-Green, Donkin, McKay, Plesken, Mann, Zelmanov, Shalev ([17], [6], [15], [16], [20], [21], [19], [23], [22]). As the analysis of conjectures has shown the main of these are as follows:

> Conjecture C: Pro-p groups of finite coclass are soluble.
>
> Conjecture A: There exists a function $f(p,r)$ such that any p-group of coclass r has a normal subgroup of class at most 2 (1 if $p = 2$) and index at most $f(p,r)$.

Conjecture C was firstly proved for p-adic analytic pro-p groups of finite coclass ($p > 3$) by Donkin [6]. The proof used the classification of p-adic simple Lie algebras and results on p-adic Chevalley groups. After that Leedham-Green showed that any pro-p group of finite coclass is p-adic analytic and thus confirmed Conjecture C for $p > 3$ [15]. Conjecture A was reduced to Conjecture C by Leedham-Green [16]. The proof is complicated and does not admit any bounds.

A constructive approach to coclass conjectures offered by Shalev and Zelmanov uses finite-dimensional modular Lie algebras instead of Lie algebras of characteristic zero. At first they obtained a short proof of Conjecture C for all primes p [23]. Soon afterwards Shalev proved Conjecture A with explicit bounds for all primes p [22].

---

*The authors were supported by International Science Foundation Grant MCN000 and Russian Foundation of Fundamental Investigations Grants N 94–01–01543 and N 94–01–00431–a.

We briefly describe here how modular Lie algebras appear in the proof of Conjecture A given by Shalev. First he constructs a certain Lie algebra M over $\mathbb{F}_p$ on which the standard graded ring $S = \bigoplus_{i\geq 1} \gamma_i/\gamma_{i+1}$ associated with a p-group G acts by derivations. Here $\gamma_i = \gamma_i(G)$ is the lower central series of group G (see, e.g. [7]). More precisely, he considers the Lie algebra $R = S/pS = \bigoplus_{i\geq 1} R_i$ and the subspace $M = \bigoplus_{i\geq m} R_i$ for some m. Let l be the maximal integer such that $(\gamma_i, \gamma_j) \subseteq \gamma_{i+j+l}$ for all $i, j \geq m$. Then the multiplication on M is defined as follows

$$[x\gamma_{i+1}, y\gamma_{j+1}] = (x, y)\gamma_{i+j+l+1}.$$

Then the p-th power map on G is employed, giving a map $t : M \to M$, $x \mapsto x^p$. It is proved that M is a Lie algebra over $\mathbb{F}_p[t]$, being simultaneously a $R \bigotimes \mathbb{F}_p[t]$-module. Further, M may be covered with a Lie algebra $\tilde{M}$ which is a free module over $\mathbb{F}_p[t]$ and on which R_1 acts by derivations as well. Finally, extending scalars, a finite-dimensional Lie algebra L over algebraic closure $\overline{\mathbb{F}_p[t]}$ is obtained,

$$L = \tilde{M} \bigotimes_{\mathbb{F}_p[t]} \overline{\mathbb{F}_p[t]}.$$

Since $\gamma_i^p = \gamma_{i+d}$, $i \geq m$, $d = (p-1)p^s$ for some s, Lie algebra L has a grading modulo d, $L = \bigoplus_{i\in \mathbb{Z}/d\mathbb{Z}} L_i$ and $[R_1, L_i] = L_{i+1}$.

For a counterexample G to Conjecture A Shalev proves that $[L, L] = L$ and R_1 has an element y such that y is a nonsingular derivation of L, which means $yx = 0$, $x \in L$, iff $x = 0$. Thus, for $D = y^{p^s}$ we have a grading $L = \bigoplus_{i\in\mathbb{Z}/(p-1)\mathbb{Z}} L_i$, $DL_i = L_{i+1}$. The existence of a grading modulo $p - 1$ and a nonsingular derivation D plays the crucial role in obtaining the contradiction with $L = [L, L]$, thus proving Conjecture A.

As it was communicated to the authors by E. I. Zelmanov, in a detailed study of p-groups using this approach there appear finite-dimensional modular Lie algebras with grading modulo $p^k - 1$, $p = \operatorname{char} F$, $L = \bigoplus_{i\in\mathbb{Z}/(p^k-1)\mathbb{Z}} L_i$ with a nonsingular derivation D such that $DL_i = L_{i+1}$. Thus there arose a problem of description of finite-dimensional modular Lie algebras with concordant nonsingular derivation and some grading modulo $p^k - 1$.

In the present paper we describe the results obtained in this direction.

2. General results

It should be mentioned that Jacobson was the first to study such Lie algebras. In 1955 he proved the following theorem [8]:

Theorem 1. *Let L be a finite-dimensional Lie algebra over an algebraically closed field F admitting a nonsingular derivation.*

i) *If* $\operatorname{char} F = 0$ *then L is nilpotent.*

ii) *If* $\operatorname{char} F = p > 0$ *and L is a Lie p-algebra then L is nilpotent.*

The proof is obtained by simple employment of the theorem on weekly closed nilsubsets. The case of a field F of positive characteristic is quite different from characteristic zero in such a respect that there exist simple Lie algebras with a nonsingulur derivation.

Example 1. Let L be a three-dimensional simple Lie algebra over $\mathbb{F}_2$ with a basis $\{a, b, c\}$ and multiplication defined as follows:

$$[a, b] = b, \quad [a, c] = c, \quad [b, c] = a.$$

The map D sending a to b, b to c, c to a is a nonsingular derivation of L.

Example 2. Let $q = 2^n$,

$$L = W(1 : \mathbf{n}) = \langle\ \mathbf{e}_\alpha,\ \ \alpha \in \mathbb{F}_{\mathbf{q}}^*\ \rangle$$

be a Zassenhaus algebra over a field F of characteristic 2 with multiplication

$$[e_\alpha, e_\beta] = (\beta - \alpha)e_{\alpha+\beta}.$$

The map $D : e_\alpha \mapsto \alpha e_\alpha$ is a nonsingular derivation and $D^{q-1} = 1$.

A derivation D of a Lie algebra L is called *periodic* if $D^N = 1$. The importance of periodic derivations is demonstrated by the theorem given below.

Theorem 2. *Let L be a finite-dimensional Lie algebra over an algebraically closed field F,* char $F = p > 0$. *If L admits a nonsingular derivation then L admits a periodic one D, $D^N = 1$, where $N = p^m - 1$ and m is the maximal dimension of tori in* Der L.

Proof. According to [1] there exists a torus T of maximal dimension in Der L such that $C_L(T) = 0$. Let $\{D_1, \ldots, D_m\}$ be a basis of T consisting of toroidal elements D_i, $D_i^p = D_i$. Let $\epsilon_1, \ldots, \epsilon_m$ be a basis of $\mathbb{F}_q \subset F$, $q = p^m$, over $\mathbb{F}_p$ and $D = \epsilon_1 D_1 + \cdots + \epsilon_m D_m$. Then $\{D, D^p, \ldots, D^{p^{m-1}}\}$ is a basis of T. Therefore, D is a nonsingular derivation of L and $D^{p^m-1} = 1$. □

The following theorem obtained in [10] specifies the result of Jacobson and shows that almost always Lie algebras with a periodic derivation are abelian.

Theorem 3. *Let L be a Lie algebra (not necessarily finite-dimensional) with a periodic derivation D, $D^N = 1$.*

i) *If* char $F = 0$ *and $N \not\equiv 0(6)$ then L is abelian.*

ii) *If* char $F = p > 0$ *and $\Delta_N = \prod_{i=0}^{N-1}((1+\epsilon^i)^N - 1) \not\equiv 0(p)$ where ϵ is a primitive N-th root of 1 in F then L is abelian.*

(Note that in the last case we can assume $N \not\equiv 0(p)$.)

Consider some examples. Calculations (with using Maple V) show that

$$\Delta_2 = -3,\ \Delta_3 = 2^2 \cdot 7,\ \Delta_4 = -3 \cdot 5^3,\ \Delta_5 = 11^2 \cdot 31,$$

$$\Delta_7 = 2^6 \cdot 29^2 \cdot 127, \ \Delta_8 = -3^7 \cdot 5^3 \cdot 17^3,$$

$$\Delta_9 = 2^2 \cdot 7 \cdot 19^4 \cdot 37^2 \cdot 73, \ \Delta_{10} = -3 \cdot 11^9 \cdot 31^3.$$

From here all exceptional characteristics p for $N = 2, 3, \dots, 10$ can be seen[1]. It may be shown that for

$N = 2, \ p = 3 \ : \ L^3 = 0;$
$N = 3, \ p = 7 \ : \ L^3 = 0;$
$N = 3, \ p = 2 \ :$ there exists a simple Lie algebra $W(1:2)$.

The fact that a Lie algebra L with derivation $D, \ D^5 = 1$, is abelian for $p = 2, 3, 5, 7$ (in general $p \neq 11, 31$) seems at first surprising.

It is worth noting some parallel between periodic derivations and regular (fixed-point-free) automorphisms. According to Kreknin's theorem [12] if $\varphi : L \longrightarrow L$ is a regular automorphism, $\varphi^N = 1$ then L is soluble with derived length $r < 2^{N-1}$, and if N is prime then L is nilpotent of class $c < (N-1)2^{N-1}$ (conjecture: $c < \frac{1}{4}(N^2 - 1)$).

However, for derivations the situation seems to be different. If we consider a simple Lie algebra L with a perodic derivation $D, \ D^N = 1$, then according to an idea of Zelmanov we can construct a Lie algebra $\tilde{L} = L \bigotimes_F \mathfrak{m}$, where $\mathfrak{m}$ is the maximal ideal of the algebra $F[t]/(t^M)$. Then $\tilde{L}$ is nilpotent of however large class M with nonsingular derivation $\tilde{D} = D \otimes 1 : x \otimes f \mapsto D(x) \otimes f, \ x \in L, \ f \in \mathfrak{m}$. When $M = 2^r$ we have a soluble Lie algebra of arbitrary class r.

The following modification of one of Kreknin's examples shows that there also exist soluble but not nilpotent Lie algebras with nonsingular derivations. Namely, let $L = \langle \, e, f_0, f_1, \dots, f_{mp-1} \, \rangle_F$ be a Lie algebra of dimension $mp + 1$ with multiplication $[e, f_k] = f_{k+1}, \ k = 0, 1, \dots, mp-1$, where $f_{mp} = f_0, \ [f_i, f_j] = 0$. Obviously, $L^{(2)} = [[L, L], [L, L]] = 0$ but $L^n \neq 0$ for any n. Consider the subfield $\mathbb{F}_q \subset F, \ q = p^2$, and an element $\lambda \in \mathbb{F}_q \setminus \mathbb{F}_p$. Then the map $D : e \mapsto \lambda e, \ f_k \mapsto (1 + k\lambda) f_k$ is a derivation of L and $D^{q-1} = 1$.

3. Simple graded Lie algebras with a nonsingular derivation

In the joint work [1] were found all simple finite-dimensional Lie algebras L of characteristic $p > 7$ admitting a nonsingular derivation. It turns out that either L is isomorphic to a special Lie algebra of Cartan type $S(m : \mathbf{n}, \omega), \ m > 2$,

$$\omega = (1 - x_1^{(p^{n_1}-1)} \dots x_m^{(p^{n_m}-1)}) dx_1 \wedge \dots \wedge dx_m,$$

[1] It is interesting to note that a circulant Δ_N was offered by Wendt [25] in connection with Fermat's last theorem.

or to a Hamiltonian Lie algebra $H(m:\mathbf{n},\omega)$, $m=2s$ and

$$\omega=\sum_{i=1}^{s} dx_i\wedge dx_{i+s}+\sum_{i,j=1}^{m}\alpha_{ij}x_i^{(p^{n_i}-1)}x_j^{(p^{n_j}-1)}dx_i\wedge dx_j, \tag{1}$$

$$\alpha_{ij}\in F,\ \det(\alpha_{ij})\neq 0.$$

These algebras are realized as Lie algebras of special derivations on a divided power algebra $A(m:\mathbf{n})$ with a basis X^α, $X^\alpha=x_1^{(\alpha_1)}\dots x_m^{(\alpha_m)}$, $0\le\alpha_i<p^{n_i}, \mathbf{n}=(n_1,\dots,n_m)$, and with multiplication

$$X^\alpha X^\beta=\binom{\alpha+\beta}{\alpha}X^{\alpha+\beta}.$$

If ∂_i is the partial derivation, $\partial_i X^{(\alpha)}=X^{(\alpha-\epsilon_i)}$, then $D=\sum_{i=1}^m f_i\partial_i\in\tilde S(m:\mathbf{n},\omega)$ iff $D\omega=0$. Here $f_i\in A(m:\mathbf{n})$. Now $S(m:\mathbf{n},\omega)=[\tilde S,\tilde S]$. The algebra $H(m:\mathbf{n},\omega)$ is defined analogously [9], [11]. Moreover, all known simple Lie algebras of characteristic $p>3$ were considered and the same result was obtained for them.

As it has been noted above, the case when a Lie algebra L over a field F of characteristic p admits a grading $\bmod(p^k-1)$ and nonsingular derivation D such that $DL_i=L_{i+1}$, $i\in\mathbb{Z}/(p^k-1)\mathbb{Z}$ is of utmost interest for the theory of p-groups.

In this direction we have received the following results.

Theorem 4. ([10]). *Let L be a simple Lie algebra over an algebraically closed field F of characteristic $p>7$ with $\mathbb{Z}/(p^k-1)\mathbb{Z}$-grading and a nonsingular derivation D, $DL_i=L_{i+1}$. If $\dim L_i=1$ then $L\cong H(2:\mathbf{n},\omega)$,*

$$\omega=(1+x_1^{(p^{n_1}-1)}x_2^{(p^{n_2}-1)})\,dx_1\wedge dx_2,\quad k=|\mathbf{n}|=n_1+n_2.$$

To construct concordant grading modulo p^k-1 and a nonsingular derivation D in an algebra L of the type $S(m:\mathbf{n},\omega)$ or $H(m:\mathbf{n},\omega)$ make use of one-one correspondence between the $\mathbb{Z}/(p^k-1)\mathbb{Z}$-gradings of L and the automorphisms $\varphi\in\operatorname{Aut}L$ of order p^k-1. Namely, let ϵ be a primitive root of 1 of degree p^k-1. The eigenspaces L_i, $L_i=\{l\in L\mid\varphi(l)=\epsilon^i l\}$, of an automorphism φ, $\varphi^{p^k-1}=1$, are the homogeneous spaces of a grading modulo p^k-1 in L. A derivation D of L is coordinated with this grading iff $\varphi D=\epsilon D\varphi$.

Consider a Lie algebra L of the type $S(m:\mathbf{n},\omega)$, $\dim L=(n-1)(p^n-1)$, where $n=n_1+\dots+n_m$. Denote $x_i^{(p^{n_i}-1)}$ by $\overline{x}_i$ and $\overline{x}_1\dots\overline{x}_m$ by e. As it was shown in [1] L admits a periodic nonsingular derivation

$$D=(1+e)\partial_1+\overline{x}_1\partial_2+\overline{x}_1\overline{x}_2\partial_3+\dots+\overline{x}_1\dots\overline{x}_{m-1}\partial_m$$

of period p^n-1. Here $D\in N_{W(m:\mathbf{n})}S(m:\mathbf{n},\omega)$ where $W(m:\mathbf{n})$ is the Lie algebra of all special derivations of $A(m:\mathbf{n})$.

Let ϵ be a primitive root of 1 of degree p^n-1, φ be an automorphism of L induced by a linear automorphism denoted by φ as well:

$$\varphi(x_1)=\epsilon^{-1}x_1,\quad\varphi(x_{i+1})=\epsilon^{-p^{n_1+\dots+n_i}}x_{i+1},\quad i=1,\dots,m-1.$$

The derivation D has weight 1 with respect to φ. Therefore, D is coordinated with the grading of L modulo $p^n - 1$ induced by automorphism φ.

To consider a Lie algebra $H(m : \mathbf{n}, \omega)$ in such a way we need some additional information on tori in $\operatorname{Der} W(m : \mathbf{n})$.

4. Tori of general position in $\operatorname{Der} W(m : \mathbf{n})$

Tori in the Lie p-algebras of Cartan type were investigated by Demushkin [3], [4]. Recall that

$$\operatorname{Der} W(m : \mathbf{n}) \cong N_{\operatorname{Der} A(m:\mathbf{n})} W(m : \mathbf{n}) = \overline{W(m : \mathbf{n})},$$

where $\overline{W(m : \mathbf{n})}$ is the p-closure of $W(m : \mathbf{n})$ in $\operatorname{Der} A(m : \mathbf{n}) = W_n$, $n = |\mathbf{n}| = n_1 + \cdots + n_m$, [27]. According to [3] the maximal dimension of tori in W_n is equal to n. Here we construct maximal tori T in $\operatorname{Der} W(m : \mathbf{n})$ such that $T \cap W(m : \mathbf{n})_0 = \{0\}$ where $W(m : \mathbf{n})_0$ is a subalgebra of $W(m : \mathbf{n})$ consisting of vector fields $D = \sum f_i \partial_i$, $f_i \in \mathfrak{m}$, $\mathfrak{m}$ is the unique maximal ideal in $A(m : \mathbf{n})$. By definition such tori are of general position.

Further, for a module M over $A(m : \mathbf{n})$ and $v \in M$, by $v(0)$ we denote the class of v in $M/\mathfrak{m}M$. Thus, $D \in W(m : \mathbf{n})_0$ iff $D(0) = 0$. Let $\Omega(m : \mathbf{n})$ be the de Rham complex over the algebra $A(m : \mathbf{n})$ associated with $W(m : \mathbf{n})$ (see, e.g. [11]), $\Omega(m : \mathbf{n}) = \bigoplus_{i=0}^{m} \Omega^i(m : \mathbf{n})$, $Z^i(m : \mathbf{n})$, $B^i(m : \mathbf{n})$, $H^i(m : \mathbf{n})$ be the corresponding spaces of cocycles, coboundaries, cohomology of dimension i. Below we denote the cohomology class of a closed form ω by $[\omega]$.

Theorem 5. *There exists a one-one correspondence between the maximal tori of general position in* $\operatorname{Der} W(m : \mathbf{n})$ *and the subspaces* Σ *in* $Z^1(m : \mathbf{n})$ *such that*

$$\dim \Sigma = m, \quad \Sigma(0) = \Omega^1(m : \mathbf{n})(0), \quad [\Sigma] = H^1(m : \mathbf{n}).$$

Proof. Let T be a maximal torus of general position in $\operatorname{Der} W(m : \mathbf{n})$. It may be shown that $\dim T = n = |\mathbf{n}| = n_1 + \cdots + n_m$. For this purpose we consider the algebra $A(m : \mathbf{n})^T$ of T-invariants in $A(m : \mathbf{n})$. If $\dim T = s$, then $A(m : \mathbf{n})^T$ has dimension p^{n-s}. It is easy to show that $C_{W(m:\mathbf{n})}(T)$ generates the $A(m : \mathbf{n})$-module $W(m : \mathbf{n})$. Consider the Lie algebra $M = A(m : \mathbf{n})^T C_{W(m:\mathbf{n})}(T)$ and its natural representation ρ on $A(m : \mathbf{n})^T$. As $W(m : \mathbf{n})$ is a TI-destribution in $\operatorname{Der} A(m : \mathbf{n})$ (see [13]), then $\rho(M)$ is a TI-destribution over $A(m : \mathbf{n})^T$. According to [13], $\rho(M) \cong W(k : \mathbf{r})$ for some $k \geq 1$, $\mathbf{r} = (r_1, \ldots, r_k)$. Pick an element D in M such that $\rho(D)$ is a semisimple element in $\overline{W(k : \mathbf{r})} \setminus W(k : \mathbf{r})_0$. Then the semisimple part D_s of the element D is contained in $\overline{W(m : \mathbf{n})} \cap C_{W_n}(T)$ and the p-closure $\langle D_s, T \rangle$ is a torus T_1 in $\overline{W(m : \mathbf{n})} \setminus W(m : \mathbf{n})_0$ such that $T \subset T_1$, $T \neq T_1$, which is impossible, for T is maximal. Thus, $\dim T = n$.

It follows that $\dim T \cap W(m : \mathbf{n}) = m$. Therefore, there exists a subspace Σ in $\Omega^1 = \operatorname{Hom}_{A(m:n)}(W(m : \mathbf{n}), A(m : \mathbf{n}))$ dual to T. Obviously, $\Sigma(0) =$

$\Omega^1(m:\mathbf{n})(0)$. Since $T\cap W(m:\mathbf{n})$ generates the $A(m:\mathbf{n})$-module $W(m:\mathbf{n})$ and $d\omega(D_1,D_2)=0$ for $\omega\in\Sigma$, $D_1,D_2\in T\cap W(m:\mathbf{n})$, then $d\omega=0$, that is $\Sigma\subseteq Z^1(m:\mathbf{n})$. Suppose that $\omega\in\Sigma$ and $\omega=df$ for some $f\in A(m:\mathbf{n})$. Then $D\omega=dD(f)=0$ for any $D\in T\cap W(m:\mathbf{n})$. Therefore, $Df=\text{const}\in F$, $D^2f=0$. As D is a semisimple element, we obtain $Df=0$ for any $D\in T\cap W(m:\mathbf{n})$. Therefore, $f=\text{const}$ and $\omega=df=0$. Thus, $[\Sigma]=H^1(m:\mathbf{n})$.

Conversely, for Σ as in the theorem let $V\subset W(m:\mathbf{n})$ be the subspace dual to Σ. It may be shown that the p-closure $\overline{V}$ in $\overline{W(m:\mathbf{n})}$ is a maximal n-dimensional torus of general position in $\overline{W(m:\mathbf{n})}$. □

5. Lie algebras of the type $H(m:\mathrm{n},\omega)$

We show that a Lie algebra of a type $H(m:\mathbf{n},\omega)$ corresponding to some form ω (1) is a Block algebra of dimension p^n-1, $n=n_1+\cdots+n_m$ [2]. Construct an algebra $H(m:\mathbf{n},\omega)$ using a torus T of general position in $\operatorname{Der}W(m:\mathbf{n})$, $m=2s$. Define T by means of a subspace $\Sigma\subset Z^1(m:\mathbf{n})$ (see theorem 5). Since $[x_i^{(p^{n_i}-1)}dx_i]$, $i=1,\dots,m$ form a basis of $H^1(m:\mathbf{n})$, a basis of Σ may be chosen as

$$\omega_i=dx_i+df_i+l_i,\quad l_i=\sum_j a_{ij}x_j^{(p^{n_j}-1)}dx_j,\quad i=1,\dots,m,$$

$$f_i\in\langle\, X^{(\alpha)},\ |\alpha|=\alpha_1+\cdots+\alpha_m\geq 2\,\rangle >,\quad a_{ij}\in F,$$

where forms l_i are linearly independent over F.

According to [14], [24] the orbit of a Hamiltonian form ω with respect to the group of admissible automorphisms of the algebra $A(m:\mathbf{n})$ is determined by the pair $(\omega(0),[\omega])$, where

$$[\omega]\in H^2(m:\mathbf{n})=\langle\,[x_i^{(p^{n_i}-1)}x_j^{(p^{n_j}-1)}dx_i\wedge dx_j],\ 1\leq i<j\leq m\,\rangle.$$

Thus, the forms ω_i (i.e. the subspace Σ) can be chosen in such a way that the Hamiltonian form $\omega=\sum_{i=1}^{s}\omega_i\wedge\omega_{i+s}$ could be written as follows

$$\omega=\sum_{i=1}^{s}dx_i\wedge dx_{i+s}+\sum_{i,j=1}^{m}a_{ij}x_i^{(p^{n_i}-1)}x_j^{(p^{n_j}-1)}dx_i\wedge dx_j,\quad a_{ij}\in F,$$

$$\det(a_{ij})\neq 0.$$

Let $D_1,\dots,D_m$ be the basis of $W(m:\mathbf{n})$ dual to the basis $\omega_1,\dots,\omega_m$ of Σ. According to the theorem the p-closure $T=\overline{\langle D_1,\dots,D_m\rangle}$ is a torus of general position in $\overline{W(m:\mathbf{n})}$. Standard considerations show that the Lie algebra $H(m:\mathbf{n},\omega)$ will be isomorphic to the algebra of Hamiltonians $f\in A(m:\mathbf{n})$ modulo constants with the Poisson bracket

$$\{f,g\}=\sum_{i=1}^{s}D_i(f)D_{i+s}(g)-D_i(g)D_{i+s}(f). \tag{2}$$

The root spaces of $H(m : \mathbf{n}, \omega)$ with respect to torus T are one-dimensional, the root system Δ is the set of nonzero elements of an elementary abelian group G of rank $n = n_1 + \cdots + n_m$.

Choose the basis of $H(m : \mathbf{n}, \omega) = A(m : \mathbf{n})/F \cdot 1$ consisting of eigenvectors v_α, $\alpha \in \Delta$. Then, according to (2),

$$[v_\alpha, v_\beta] = \langle \alpha, \beta \rangle \, v_{\alpha+\beta},$$

where $\langle \ , \ \rangle$ is a nondegenerate skewsymmetric bilinear form on G with values in the field F. Thus, the Lie algebra $H(m : \mathbf{n}, \omega)$ corresponding to any form ω (1) is a Block algebra of dimension $p^n - 1$ [2]. Note that, as it follows from (2), any skewsymmetric nondegenerate form on G can be obtained in such a way. The fact that any Block algebra belongs to the class of Hamiltonian Lie algebras was obtained by Wilson [26]. Thus, we have proved the theorem.

Theorem 6. *The class of Hamiltonian Lie algebras corresponding to the forms* (1) *coincides with the class of Block algebras of dimension* $p^n - 1$.

Identify the group G with the additive group of the finite field $\mathbb{F}_q \subset F$, $q = p^n$. A $\mathbb{F}_p$-bilinear form $\langle \ , \ \rangle$ on G, in general, is $\mathbb{F}_l$-bilinear for some $l = p^t$, $t | n$. A linear map D on $H(m : \mathbf{n}, \omega)$ such that $D(v_\alpha) = \alpha v_\alpha$ is a periodic derivation of the Lie algebra $H(m : \mathbf{n}, \omega)$ of period $p^n - 1$. Let ϵ be a primitive root of 1 of degree $p^t - 1$ in the field F. The map φ, $\varphi : v_\alpha \mapsto \epsilon^{-2} v_{\epsilon\alpha}$, defines an automorphism of the algebra $H(m : \mathbf{n}, \omega)$ of order $p^t - 1$, $\varphi D \varphi^{-1} = \varphi(D) = \epsilon D$. Thus, the derivation D is coordinated with the cyclic grading induced by the automorphism φ.

Sum up the results obtained.

Theorem 7. $(p > 7)$. *A Lie algebra* $S(m : \mathbf{n}, \omega)$,

$$\omega = (1 - x_1^{(p^{n_1}-1)} \ldots x_m^{(p^{n_m}-1)}) dx_1 \wedge \cdots \wedge dx_m,$$

admits a periodic derivation of period $p^n - 1$, $n = n_1 + \cdots + n_m$, *coordinated with a cyclic grading modulo* $p^n - 1$.

A Lie algebra $H(m : \mathbf{n}, \omega)$, $m = 2s$,

$$\omega = \sum_{i=1}^{s} dx_i \wedge dx_{i+s} + \sum_{i,j=1}^{m} \alpha_{ij} x_i^{(p^{n_i}-1)} x_j^{(p^{n_j}-1)} dx_i \wedge dx_j,$$

$$\alpha_{ij} \in F, \quad \det(\alpha_{ij}) \neq 0$$

admits a periodic derivation of period $p^n - 1$, $n = n_1 + \cdots + n_m$, *coordinated with a cyclic grading modulo* $p^t - 1$ *for some* $t \geq 1$, $t | n$.

Note that the problem of maximal possible t for a given Hamiltonian Lie algebra remains open. There exist Lie algebras for which t can not be greater than 1. On the other hand, according to theorem 4 $t = n$ iff $m = 2$.

References

[1] G. Benkart, A. I. Kostrikin and M. I. Kuznetsov, Finite-dimensional simple Lie algebras with a nonsingular derivation, J. Algebra (to appear).

[2] R. E. Block, New simple Lie algebras of prime characteristic, Trans. Amer. Math. Soc. 89 (1958), 421–449.

[3] S. P. Demushkin, Cartan subalgebras of the simple Lie p-algebras W_n and S_n (Russian), Sibirsk. Mat. Z. 11 (1970), 310–325; English transl. Siberian Math. J. 11 (1970), 233–245.

[4] S. P. Demushkin, Cartan subalgebras of simple nonclassical Lie p-algebras (Russian), Izv. Akad. Nauk SSSR Ser. Mat. 36 (1972), 915-932; English transl. Math. USSR-Izv. 6 (1972), 905–924.

[5] J. Dixon, M. P. F. du Sautoy, A. Mann and D. Segal, Analytic Pro-p groups, London Math. Soc. Lecture Note Ser. no. 157, Cambridge University Press, Cambridge, 1991.

[6] S. Donkin, Space groups and groups of prime power order VIII. Pro-p groups of finite coclass and p-adic Lie algebras, J. Algebra 111 (1987), 316–342.

[7] B. Huppert and N. Blackburn, Finite Groups II, Springer-Verlag, Berlin, 1982.

[8] N. Jacobson, A note on automorphisms and derivations of Lie algebras, Proc. Amer. Math. Soc. 6 (1955), 281–283.

[9] V. G. Kac, Description of filtered Lie algebras with which graded Lie algebras of Cartan type are associated (Russian), Izv. Akad. Nauk SSSR Ser. Mat. 38 (1974), 800–834; English transl. Math. USSR-Izv. 8 (1974), 801–835.

[10] A. I. Kostrikin and M. I. Kuznetsov, Two notes on Lie algebras with a nonsingular derivation (Russian), Proc. Steklov Inst. Math. (to appear).

[11] A. I. Kostrikin and I. R. Shafarevich, Graded Lie algebras of finite characteristic (Russian), Izv. Akad. Nauk SSSR Ser. Mat. 33 (1969), 251–322; English transl. Math. USSR-Izv. 3 (1969), 237–304.

[12] V. A. Kreknin, Solvability of Lie algebras with a regular automorphism of finite period (Russian), Dokl. Akad. Nauk SSSR, 150 (1963), 467–469; English transl. Soviet Math. Dokl. 4 (1963), 683–685.

[13] M. I. Kuznetsov, Destributions over an algebra of truncated polynomials (Russian), Matem. Sbornik 136 (1988), 187–205; English transl. Math. USSR Sbornik 64 (1989), 187–205.

[14] M. I. Kuznetsov and S. A. Kirillov, Hamiltonian differential forms over an algebra of truncated polynomials (Russian), Uspehi Mat. Nauk 41 (1986), 205–206.

[15] C. R. Leedham-Green, Pro-p groups of finite coclass, Preprint.

[16] C. R. Leedham-Green, The structure of finite p-groups, Preprint.

[17] C. R. Leedham-Green, S. McKay and W. Plesken, Space groups and groups of prime power order, V. A bound to the dimension of space groups with fixed coclass, Proc. London Math. Soc. 52 (1986), 73–94.

[18] C. R. Leedham-Green and M. F. Newman, Space groups and groups of prime power order I, Arch. Math. 35 (1980), 193–202.

[19] A. Mann, Space groups and groups of prime power order VII. Powerfull p-groups and uncovered p-groups, Bull. London Math. Soc. (to appear).

[20] S. McKay, On the structure of a special class of p-groups, Quart. J. Math. Oxford Ser. (2) 38 (1987), 489–502.

[21] S. McKay, On the structure of a special class of p-groups II, Quart. J. Math. Oxford Ser. (2) 41 (1990), 431–448.

[22] A. Shalev, The structure of finite p-groups: constructive proof of coclass conjectures, Preprint.

[23] A. Shalev and E. I. Zelmanov, Pro-p groups of finite coclass, Math. Proc. Camb. Phil. Soc. 111 (1992), 417–421.

[24] S. M. Skryabin, Classification of Hamiltonian forms over divided power algebras (Russian), Matem. Sbornik 181 (1990), 114–133.

[25] E. Wendt, Arithmetischen Studien über den letzten Fermatschen Satz, welcher aussagt daß die Gleichung $a^n = b^n + c^n$ für $n > 2$, in ganzen Zahlen nicht auflösbar ist, J. reine und angew. Math. 113 (1894), 335–346.

[26] R. L. Wilson, Nonclassical simple Lie algebras, Bull. Amer. Math. Soc. 75 (1969), 987–991.

[27] R. L. Wilson, Classification of generalized Witt algebras over algebraically closed fields, Trans. Amer. Math. Soc. 153 (1971), 191–210.

The irregularity of cyclic multiple plane curves

*Vic. S. Kulikov**

0.

Let $D \subset \mathbb{C}^2$ be a complex algebraic curve defined by an equation $f(x,y) = 0$, where $f(x,y) \in \mathbb{C}[x,y]$, and

$$f(x,y) = \prod_{i=1}^{k} f_i^{m_i}(x,y) \tag{1}$$

be the decomposition into irreducible factors. Denote by $D_i \subset \mathbb{C}^2$ an irreducible component of the curve D defined by $f_i(x,y) = 0$. The equation $z = f(x,y)$ defines a morphism $\phi : \mathbb{C}^2 \to \mathbb{C}^1$. We assume that the generic fibre of this morphism is irreducible. In particular, we assume that $\mathrm{GCD}(m_1, \ldots, m_k) = 1$.

Let $F_n : X_n \to \mathbb{C}^2$ be the normalization of a cyclic n-sheeted covering ramified over D and defined in $\mathbb{C}^3$ by the equation $z^n = f(x,y)$. Put $X_n^0 = F_n^{-1}(\mathbb{C}^2 \setminus D)$.

Let $\mathbb{C}^2 \subset \mathbb{P}^2$ be the standard compactification associated with a linear structure in $\mathbb{C}^2$ defined by coordinates (x,y) in $\mathbb{C}^2$ (i.e. if $(x_0 : x_1 : x_2)$ are the homogeneous coordinates in $\mathbb{P}^2$, then $x = \frac{x_1}{x_0}$, $y = \frac{x_2}{x_0}$). Denote by $\bar{D}_i$ the closure of $D_i \subset \mathbb{C}^2$ in $\mathbb{P}^2$, and $L_\infty = \mathbb{P}^2 \setminus \mathbb{C}^2$ the line at infinity. Let $\bar{f}(x_0, x_1, x_2)$ be the homogeneous polynomial associated with f. Let $\bar{X}_n$ denote the normalization of the projective closure of X_n, i.e. $\bar{X}_n$ is the normalization of the surface in $\mathbb{P}^3$ defined by

$$x_3^n = x_0^m \bar{f}(x_0, x_1, x_2), \tag{2}$$

where $m = n - \deg f$. Let $\tilde{X}_n$ be a desingularization of $\bar{X}_n$, $\bar{F}_n : \bar{X}_n \to \mathbb{P}^2$ and $\tilde{F}_n : \tilde{X}_n \to \mathbb{P}^2$ be the corresponding morphisms induced by F_n. Let $B_i = F_n^{-1}(D_i)$ be the strict pre-image of D_i.

The morphism ϕ induces an epimorphism

$$\phi_* : \pi_1(\mathbb{C}^2 \setminus D) \to \pi_1(\mathbb{C}^1 \setminus \{0\}) \cong \mathbb{F}_1,$$

where $\mathbb{F}_1$ is the free cyclic group. Denote by N the kernel of ϕ_*. The epimorphism ϕ_* defines an infinite cyclic covering $F_\infty : X_\infty \to \mathbb{C}^2 \setminus D$ such that $\pi_1(X_\infty) = N$

*The research described in this publication was made possible in part by Grant No. 94-01-01575 from Russian Foundation of Fundamental Researches.

and $H_1(X_\infty, \mathbb{Z}) = N/N'$, where $N' = [N, N]$ is the commutator-group of N. For each n there is a holomorpic mapping $F_{\infty,n} : X_\infty \to X_n^0$ such that the following diagram is commutative

$$\begin{array}{ccc} X_\infty & \xrightarrow{F_{\infty,n}} & X_n^0 \\ & \searrow^{F_\infty} & \downarrow F_n \\ & & \mathbb{C}^2 \setminus D. \end{array}$$

If the generic fibre of the morphism ϕ is irreducible, then [5] N is a finitely generated group.

The exact sequence

$$1 \to N/N' \to \pi_1(\mathbb{C}^2 \setminus D)/N' \to \mathbb{F}_1 \to 1$$

defines an action of a generator t of the free group $\mathbb{F}_1$ on N/N' and defines a structure of $\mathbb{Z}[t, t^{-1}]$-module on N/N'. Let $\Delta_{i,D}(t) = \Delta_i(t)$ be the i-th Alexander polynomial associated with the action of t on N/N'. The definition of the Alexander polynomials and their properties can be founded in [4, 5]. We note here only that $\Delta_1(t)$ coincides with the characteristic polynomial of the automorphism $h = h_D \in \mathrm{Aut}(N/N' \otimes \mathbb{C})$ induced by the action of the generator $t \in \mathbb{F}_1$ on N/N'. We call h_D the Alexander automorphism associated with the curve D. Let $e_j(t)$, $j = 1, \ldots, \dim(N/N' \otimes \mathbb{C})$, be the elementary divisors of $h - t \cdot \mathrm{Id}$. It is well known [4, 5], that the roots of the Alexander polynomial $\Delta_1(t)$ are roots of unity. Let $n(D)$ be the minimal natural number such that all roots of $\Delta_1(t)$ are $n(D)$-th roots of unity.

Denote by $R(k, D)$ the union of sets of distinct roots of $e_j(t)$, which are k-th roots of unity unequal to 1, and put

$$N(k, D) = \#R(k, D)$$

Note that $N(k, D)$ coincides with the number of Jordan blocks of the automorphism h corresponding to eigen-values belonging to $R(k, D)$.

We say that a curve $D = m_1 D_1 + \cdots + m_k D_k$ is *connected modulo n*, if

$$D_{\mathrm{mod}\, n} = \sum_{m_i \not\equiv 0 \bmod n} m_i D_i$$

is connected in $\mathbb{C}^2$. A curve $D = m_1 D_1 + \cdots + m_k D_k$ is called *absolutely connected modulo n* if D is connected modulo n_1 for each divisor $n_1 > 1$ of n.

The aim of this paper is to calculate the irregularity

$$q(\tilde{X}_n) = \dim H^0(\tilde{X}_n, \Omega^1_{\tilde{X}_n}) = \frac{1}{2} \dim H^1(\tilde{X}_n, \mathbb{R})$$

in terms of the number of n-th roots of unity which are the roots of $\Delta_{1,D}(t)$ in the case D is absolutely connected modulo $n(D)$.

The relations between homology of X_n^0 and $\tilde{X}_n$ when D is irreducible and has "a good behaviour" at infinity were investigated in [9, 6, 8, 3]. In particular, A. Libgober proved the following

Theorem ([6]). *Let $D \subset \mathbb{C}^2$ be an irreducible curve such that*

$$\mathrm{GCD}(t_1, \dots, t_D, n) = 1,$$

where $t_1, \dots, t_D$ are the intersection indices of $\bar{D}$ and the line at infinity L_∞. Then

i) *the dimension of the kernel of the epimorphism*

$$i_* : H_1(X_n^0, \mathbb{C}) \to H_1(\tilde{X}_n, \mathbb{C})$$

induced by inclusion $i : X_n^0 \hookrightarrow \tilde{X}_n$ is equal to $\dim\ker i_* = 1$;

ii) $\dim H_1(\tilde{X}_n, \mathbb{C}) = N(n, D)$.

The main result of this paper is the following theorem

Theorem 1. *Let $D = m_1D_1 + \cdots + m_kD_k$ be absolutely connected modulo n. Then*

$$k \leq \dim\ker i_* \leq \#\{\textit{irreducible components of the divisor}\, F_n^{-1}(D),\}$$

where the epimorphism $i_* : H_1(X_n^0, \mathbb{C}) \to H_1(\tilde{X}_n, \mathbb{C})$ is induced by the inclusion $i : X_n^0 \hookrightarrow \tilde{X}_n$.

Corollary 1. *Let $D = m_1D_1 + \cdots + m_kD_k$ be absolutely connected modulo n and let all curves $B_i = F_n^{-1}(D_i)$, $1 \leq i \leq k$, be irreducible. Then* $\dim\ker i_* = k$.

Corollary 2. *If D is connected and every multiplicity m_i is relatively prime with n (in particular, if D is reduced, i.e. all multiplicities m_i are equal to 1), then* $\dim\ker i_* = k$.

Corollary 3. *If D is irreducible, then* $\dim\ker i_* = 1$.

Theorem 2. *Let $D \subset \mathbb{C}^2$ be absolutely connected modulo $n(D)$ and let $B_i = F_{n(D)}^{-1}(D_i)$, $1 \leq i \leq k$, be irreducible. Then the irregularity*

$$q(\tilde{X}_n) = \dim H^0(\tilde{X}_n, \Omega^1_{\tilde{X}_n}) = \frac{1}{2}N(n, D)$$

for all n.

The proof of the Theorems 1 and 2 is based on the following

Theorem 3. *Let D be absolutely connected modulo n and let ν and ν^{-1} be regular functions on X_n. Then* $\nu = \mathrm{const}$.

1.

Proof of Theorem 3. A compactification $\mathbb{P}^2$ of $\mathbb{C}^2$ and a nonsingular projective model $\tilde{X}_n$ of n-sheeted cyclic covering X_n of the affine space $\mathbb{C}^2$ ramified along D are defined not uniquely but modulo birational isomorphisms.

Lemma 1. *Let $D \subset \mathbb{C}^2$ be any curve, $\deg D = d$, defined by (1) and $F_n : X_n \to \mathbb{C}^2$ be the cyclic covering defined by $z^n = f(x,y)$. Then there exist projective completions $\mathbb{C}^2 \subset \mathbb{P}^2$ and $X_n \subset \tilde{X}_n$ such that the morphism $\tilde{F}_n : \tilde{X}_n \to \mathbb{P}^2$ isn't ramified along the line $L_\infty = \mathbb{P}^2 \setminus \mathbb{C}^2$, and moreover the pre-image $\tilde{F}_n^{-1}(L_\infty)$ splits into n irreducible components.*

Proof. Put $\tilde{n} = d \cdot n$, where $d = \deg f(x,y)$, and let $\mathbb{C}^2 \subset \mathbb{P}^2$ be the standard compactification. Choose a straight line $L \subset \mathbb{P}^2$ such that L does not intersect $\bar{D}$ at infinity. Performing linear change of variables, we can assume that L is defined by equation $x = 0$. Then we perform the following change of variables in $\mathbb{C}^2$:

$$\begin{aligned} u &= x \\ v &= y + x^m \end{aligned}$$

It is easy to see that the curve D is defined by an equation $\tilde{f}(u,v) = 0$, where $\deg \tilde{f} = \tilde{n}$ and $\tilde{f}(u,v) = u^{\tilde{n}} + \tilde{f}_1(u,v)$ such that $\deg \tilde{f}_1 < \tilde{n}$. Thus the cyclic covering $\tilde{F}_{\tilde{n}} : \tilde{X}_{\tilde{n}} \to \mathbb{P}^2$ is not ramified along the line at infinity L_∞, where $\mathbb{P}^2$ is the standard compactification $\mathbb{C}^2$ corresponding to linear structure defined by coordinates (u,v). It is obvious that $\tilde{F}_{\tilde{n}}^{-1}(L_\infty)$ splits into n irreducible components, since L_∞ intersects the ramification divisor only at one point. In this case, the cyclic covering $\tilde{F}_n : \tilde{X}_n \to \mathbb{P}^2$ is not ramified along L_∞, because we can choose compactifications $\tilde{X}_{\tilde{n}}$ and $\tilde{X}_n$ such that $\tilde{F}_{\tilde{n}}$ and $\tilde{F}_n$ can be included in the following commutative diagram

$$\begin{array}{ccc} \tilde{X}_{\tilde{n}} & \xrightarrow{F_{\tilde{n},n}} & \tilde{X}_n \\ & \tilde{F}_{\tilde{n}} \searrow & \downarrow \tilde{F}_n \\ & & \mathbb{P}^2 \end{array}$$

where $F_{\tilde{n},n}$ is a cyclic covering of degree d. Lemma 1 is proven.

Fix a compactification $\mathbb{C}^2 \subset \mathbb{P}^2$ guaranteed by Lemma 1. Denote by I_n the set of all not constant regular funtions ν on X_n such that ν^{-1} are regular on X_n too. We need to show that $I_n = \emptyset$. Assume the contrary, i.e. $I_n \neq \emptyset$.

Denote by $h_n^* \in \mathrm{Gal}(\mathbb{C}(X_n)/\mathbb{C}(\mathbb{C}^2))$ a generator of the cyclic group of automorphisms of the rational function field $\mathbb{C}(X_n)$ of X_n over $\mathbb{C}(\mathbb{C}^2)$. The automorphism h_n^* is induced by the automorphism h_n of the manifold X_n over $\mathbb{C}^2$. We have $h_n^{*n} = \mathrm{Id}$. Thus all eigen-values of the automorphism h_n^* are n-th roots of unity.

Obviousely, I_n is invariant under the action of h_n^*. For each $\nu \in I_n$, denote by n_ν the minimal degree for which $(h_n^*)^{n_\nu}(\nu) = \lambda\nu$, $\lambda \in \mathbb{C}$. Put

$$n_{\min} = \min_{\nu \in I_n} n_\nu .$$

Note that $n_{\min} > 1$ if $I_n \neq \emptyset$. In fact, if $h_n^*(\nu) = \lambda\nu$, then $h_n^*(\nu^p) = \lambda^p\nu^p$. Thus for $p = n$, we have $h_n^*(\nu^n) = \nu^n$, since all eigen-values of the automorphism h_n^* are n-th roots of unity. Since $h_n^*(\nu^n) = \nu^n$ it follows that ν^n is lifted from $\mathbb{C}^2$. Thus

$\nu^n \in \mathbb{C}$ (and, consequently, $\nu \in \mathbb{C}$), since every regular and invertible function on $X_1 = \mathbb{C}^2$ is equal to a constant. So ν does not belong to I_n.

In any case, if a function $\nu \in I_n$ satisfies the condition $(h_n^*)^k(\nu) = \lambda\nu$ for some $\lambda \in \mathbb{C}$, then $I_k \neq \emptyset$, since in this case the function ν^n is lifted from X_k and $\nu^n \neq \text{const}$ if $\nu \neq \text{const}$, i.e. ν^n belongs to I_k. Thus, without loss of generality, we can assume that $n = n_{\min}$. In this case, if $I_n \neq \emptyset$, then

$$\frac{(h_n^*)^k(\nu)}{\nu} \in I_n$$

for each function $\nu \in I_n$ and for each k, $0 < k < n$.

Let $\sigma : \bar{\mathbb{P}}^2 \to \mathbb{P}^2$ be a composition of σ-processes with center $\bar{D} \cap L_\infty$ such that $\sigma^{-1}(D \cup L_\infty)$ is a divisor with normal crossings in some nebourhood of the pre-image of L_∞. Denote by $C_1, \dots, C_s$ the irreducible components of $\sigma^*(\bar{D} + L_\infty)$ and let C_i be the strict pre-images of the curves $\bar{D}_i$ for $i < k+1$, C_{k+1} be the strict pre-image of L_∞, and, for $i > k+1$, C_i be the exeptional curves of the morphism σ.

Let $\hat{X}_n$ be the normalization of $\bar{\mathbb{P}}^2$ in the field $\mathbb{C}(X_n)$, and $\hat{F}_n : \hat{X}_n \to \bar{\mathbb{P}}^2$ be the corresponding morphism. The divisor of the rational function f on $\bar{\mathbb{P}}^2$ determining our n-sheeted covering is equal to

$$(f) = m_1 C_1 + \cdots + m_k C_k - dC_{k+1} + \sum_{i=k+2}^{s} m_i C_i,$$

where $d = \deg f$. The morphism $\hat{F}_n$ is not ramified outside the curves C_i and by assumption is not ramified along C_{k+1}. The index of ramification of the morphism $\hat{F}_n$ along a curve C_i is equal to $n/\operatorname{GCD}(n, m_i)$. By $B_{i,j}$ denote irreducible components of the strict pre-image of $\hat{F}_n^*(C_i)$ of curves C_i. By assumption (Lemma 1) $\hat{F}_n^*(C_{k+1}) = B_{k+1,1} + \cdots + B_{k+1,n}$.

Remark 1. If $\nu \in I_n$, then the divisor

$$(\nu) = \sum_{i>k} \sum_{j=1}^{n_i} r_{i,j} B_{i,j},$$

moreover there exists j for which $r_{k+1,j} \neq 0$, since otherwise ν would be regular and invertible on $\bar{X}_n \setminus \operatorname{Sing} \bar{X}_n$, $\nu \neq \text{const}$, that is impossible, since $\bar{X}_n$ is a normal manifold.

Let Γ be a dual graph of the divisor

$$C = \sum_{i=1}^{s} C_i,$$

where vertices v_i of the graph Γ correspond to the irreducible components C_i of the divisor C, and the edges correspond to the intersection points of the component C_i and C_j of the divisor C such that either $i > k$, or $j > k$. Prescribe a multiplicity m_i to the vertex v_i, where m_i is the multiplicity of the corresponding component

C_i in the divisor (f). Let Γ_∞ be a graph corresponding to the divisor

$$C_\infty = \sum_{i=k+1}^{s} C_i.$$

Note that Γ_∞ is a tree.

We call v_i *a final vertex* if this vertex is connected in Γ with other vertex v_j such that the multiplicity m_j is not divisible by n.

Lemma 2. *If $v_i \in \Gamma$ is a final vertex, then the corresponding curve C_i satisfies the following conditions:*

i) $F_n^{-1}(C_i) = B_{i,1} + \cdots + B_{i,n_i}$, *where* $n_i < n$,

ii) *if n is a divisor of m_i, i.e. the curve C_i does not belong to the ramification divisor of the morphism $\hat{F}_n$, then there exists a number r_i, $1 \leq r_i < n$, such that $h_n^{r_i}(B_{i,j}) = B_{i,j}$ for all j, $1 \leq j \leq n_i$.*

iii) $C_i \neq C_{k+1}$.

Proof. If n is not a divisor of m_i, then the curve C_i belongs to the ramification divisor of $\hat{F}_n$ and, obviousely, satisfies the assertion of Lemma. Let C_i does not belong to the ramification divisor of $\hat{F}_n$. Then the morphism $\hat{F}_n : \hat{F}_n^{-1}(C_i) \to C_i$ is ramified, since v_i is a final vertex. Thus C_i satisfies the condition i) and, obviousely, satisfies the conditions ii) and iii). Lemma 2 is proven.

Put $E = \{i \in \{1, \ldots, s\} \mid n \text{ is a divisor of } m_i,\ v_i \text{ is a final vertex }\}$.

Lemma 3. *If $I_n \neq \emptyset$, then there exists a function $\nu_0 \in I_0$ such that for each $i \in E$ and for each j the curves $B_{i,j}$ are not components of the divisor (ν_0).*

Proof. Choose some $\nu \in I_n$ and some $i_0 \in E$. It follows from Lemma 2 that there exists $r_0 < n$ such that all $B_{i_0,j}$ are invariant under $(h^*)^{r_0}$. Since $n = n_{\min}$ and $1 \leq r_0 < n$, then the function

$$\nu' = \frac{(h_n^*)^{r_0}(\nu)}{\nu} \neq \text{const}$$

and belongs to I_n. It is easy to see that the curves $B_{i_0,j}$ are not the components of the divisor (ν'). If $E \setminus \{i_0\} \neq \emptyset$, then we choose another index i_1 and we construct a function ν'' as above using the function ν' instead of the function ν. Looking over all indices of E we obtain desired function. Lemma 3 is proven.

Remove all final vertices v_i, $i \in E$, from the graph Γ_∞. We obtain a nonempty graph one of the connected components of which contains the vertex v_{k+1}. Denote this component by Γ'_∞.

Lemma 4. *For*

$$C_{\Gamma'_\infty} = \sum_{v_i \in \Gamma'_\infty} C_i,$$

the pre-image $\hat{F}_n^{-1}(C_{\Gamma'_\infty})$ is decomposed into n connected components $B_{\Gamma'_\infty,j}$, $1 \leq j \leq n$. Moreover, for each j, the morphism $\hat{F}_n : B_{\Gamma'_\infty,j} \to C_{\Gamma'_\infty}$ is an isomorphism.

Proof. The result follows from the facts that for $v_i \in \Gamma'_\infty$ the curves C_i are rational, do not intersect the ramification divisor of the morphism $\hat{F}_n$, and the graph Γ'_∞ is a tree.

Let us consider the function ν_0, the existence of which is proven in Lemma 3 in the case $I_0 \neq \emptyset$. Without loss of generality, we can assume that ν_0 defines a regular morphism $\bar{\nu} : \hat{X}_n \to \mathbb{P}^1$.

Lemma 5. i) *The constructed above function* ν_0 *has no zeros and poles along the curves* $B_{i,j}$ *for each* v_i *not belonging to* Γ'_∞ *and for each* j.

ii) *If* ν_0 *has zeros (correspondently, poles) along* B_{k+1,j_0}, *then* ν_0 *has no poles (correspondently, zeros) along* B_{i,j_0} *for each* $v_i \in \Gamma'_\infty$.

iii) *If* ν_0 *and* ν_0^{-1} *are regular at a generic point of* B_{k+1,j_0}, *then these functions are regular along each curve* B_{i,j_0} *for each* $v_i \in \Gamma'_\infty$.

Proof. We give a proof of i) only, because proofs of ii) and iii) are similar.

By construction the functions ν_0 and ν_0^{-1} are regular along the curves $B_{i,j}$ for $i < k+1$, or if v_i is a final vertex. For $i > k+1$ such that v_i are not in Γ'_∞ and v_i are not final vertices, all curves C_i can be simultaneously contracted into singular points since the intersection matrix of these curves is negatively defined. Hence the pre-images of these curves (the curves $B_{i,j}$) can be simultaneously contracted into singular points too. Therefore Lemma 5 follows from the following well known fact. Let $p \in Z$ be a point of a normal surface (possibly noncomplete), and ν be a rational function on Z such that ν and ν^{-1} are regular on $Z \setminus p$. Then ν is regular at p and $\nu(p) \neq 0$.

As noted above, the function ν_0 should have zeros and poles along some curves $B_{k+1,j}$. For definitness assume that ν_0 has zero along $B_{k+1,1} \subset B_{\Gamma'_\infty,1}$. The automorphism h_n transposes divisors $B_{\Gamma'_\infty,j}$ and for $j_1 \neq j_2$ the divisors $B_{\Gamma'_\infty,j_1}$ and $B_{\Gamma'_\infty,j_2}$ are not intersected. Therefore h_n transposes connected components of fibres of $\bar{\nu}$, i.e. fibres of the morphisms $\bar{\nu}$ and $h_n^*(\bar{\nu})$ (defined by the function $h_n^*(\nu_0)$) have the same connected components. If fibres of the morphism $\bar{\nu}$ are not connected, then by Stein factorization theorem there exist a curve L and morphisms $\mu : L \to \mathbb{P}^1$ and $\hat{\nu} : \hat{X}_n \to L$ such that $\bar{\nu} = \mu \cdot \hat{\nu}$ and the fibres of $\hat{\nu}$ are connected.

The structure of the fibering defined by $\hat{\nu}$ is invariant under the action of the automorphism h_n. Thus the automorphism h_n determines an automorphism $\hat{h}$ of the curve L. Denote by $\phi_n : L \to L/\hat{h}$ the factorization under the action of $\hat{h}$. Then there exists a morphism $\hat{\psi} : \hat{\mathbb{P}}^2 \to L/\hat{h}$ such that the following diagram is commutative:

$$\begin{array}{ccc} \hat{X}_n & \xrightarrow{\hat{F}_n} & \hat{\mathbb{P}}^2 \\ {\scriptstyle \hat{\nu}}\downarrow & & \downarrow{\scriptstyle \hat{\psi}} \\ L & \xrightarrow{\phi_n} & L/\hat{h} \end{array} \qquad (3)$$

We can assume that the fibres of the morphism $\hat{\psi}$ are connected too. Since the divisor $\sum B_{\Gamma'_\infty,j}$ contains n distinct fibres of the morphism $\hat{\nu}$ (say $Y_1,\dots,Y_n$), then $\deg\phi_n = n$. Since $\hat{\mathbb{P}}^2$ is a rational surface, then $L/\hat{h} \cong \mathbb{P}^1$.

The morphism $\hat{\psi}\cdot\hat{F}_n$ maps all Y_j into a unique point. Denote it by ∞. Note that ϕ_n is not ramified at ∞. From Lemma 5 and from commutative diagram (3) it follows that $\operatorname{supp}\hat{\psi}^{-1}(\infty) \subset \operatorname{supp} C_{\Gamma'_\infty}$ (and without loss of generality, we can assume that $\operatorname{supp}\hat{\psi}^{-1}(\infty) = \operatorname{supp} C_{\Gamma'_\infty}$). Since for each j the image $\hat{F}_n(Y_j) = C_{\Gamma'_\infty}$ lies in the complement of $\mathbb{C}^2$, the restriction $\hat{\psi}\,|_{\mathbb{C}^2} = \psi$ maps $\mathbb{C}^2$ onto $\mathbb{P}^1\setminus\{\infty\} \cong \mathbb{C}^1$. If we choose coordinates in $\mathbb{C}^2$ and in $\mathbb{C}^1$, then the morphism ψ is determined by an equation $z = g(x,y)$, where $g(x,y)$ is a polynomial. Note that for each constant $c \in \mathbb{C}$ and for each natural number $r > 1$, the polynomial $g(x,y) - c$ is not an r-th power of some polynomial, since the fibres of the morphism ψ are connected.

For all $i < k+1$ such that n is not divisible by m_i, the curves $B_{i,j}$ are components of fibres of the morphism $\hat{\nu}$, since, by construction of ν_0, these curves do not intersect $B_{\Gamma'_\infty}$. On the other hand, the restriction of $\hat{F}_n$ on X_n coincides with F_n, which is ramified along $D_{\mathrm{mod}\,n}$ only. It follows from absolutely connectedness modulo n of D that all components of the divisor $D_{\mathrm{mod}\,n}$ should be components of a unique fibre of the morphism ψ. We denote this fibre by S_0, $\psi(S_0) = 0 \in \mathbb{C}^1 \subset \mathbb{P}^1$.

We show that the morphism $\phi_n : L \to \mathbb{P}^1$ is not ramified at all points $z \in \mathbb{P}^1 \setminus \{0\}$. In fact, as it was noted above, ϕ_n is not ramified at $\infty \in \mathbb{P}^1$. Assume that ϕ_n is ramified at a point $z = c$, $c \neq 0$. Choose a local parameter τ in the pre-image of the point $z = c$ such that $\tau^{n_1} = \phi_n^*(z - c)$, where $n_1 > 1$. We have a necessary condition for the morphism F_n to be not ramified at $\psi^{-1}(\{z = c\})$: all components of the curve defined in $\mathbb{C}^2$ by the equation $g(x,y) = c$ have multiplicities divisible by n_1. Hence, if ϕ_n is ramified at $z = c$, then we can extract the n_1-th root from the polynomial $g(x,y) - c$, that is impossible.

To conclude the proof of Theorem 3, note that a morphism $\phi_n : L \to \mathbb{P}^1$ of degree $\deg\phi_n > 1$ should have at least two ramification points.

2.

Theorem 3 yields the following

Theorem 4. *Let a curve $D \subset \mathbb{C}^2$ be absolutely connected modulo n, $F_n : X_n \to \mathbb{C}^2$ and $\tilde{F}_n : \tilde{X}_n \to \mathbb{P}^2$ be coverings determined by D. Then $i_* : H^1(X_n \setminus \operatorname{Sing} X_n, \mathbb{Q}) \to H^1(\tilde{X}_n, \mathbb{Q})$ is an isomorphism, where i_* induced by the inclusion $i : X_n \setminus \operatorname{Sing} X_n \hookrightarrow \tilde{X}_n$.*

Proof. We can assume that $\tilde{X}_n \setminus (X_n \setminus \operatorname{Sing} X_n)$ is a divisor with normal crossings. The components of this divisor are linearly independent in $\operatorname{Pic}(\tilde{X}_n) \otimes \mathbb{Q}$, since otherwise there exists a regular function $\nu \neq \mathrm{const}$ on X_n such that ν^{-1} is regular on X_n, that is impossible. Therefore Theorem 4 follows from

Proposition 1. ([2, 1]). *Let Z be a smooth projective surface and let $D = D_1 + \cdots + D_s$ be a divisor with normal crossings on Z. Then*

$$b_1(Z \setminus D) - b_1(Z) = s - \rho(D),$$

where $b_1(\cdot)$ is the first Betti number and $\rho(D) = \dim\{\sum \mathbb{Q}[D_i]\}$ in $\mathrm{Pic}(Z) \otimes \mathbb{Q}$.

Proof. One can deduce this from the Residue sequence:

$$0 \to H^0(Z, \Omega_Z^1) \to H^0(Z, \Omega_Z^1(\log D)) \to H^0(D^{(1)}, \mathcal{O}_{D^{(1)}}) \to H^1(Z, \Omega_Z^1),$$

where $D^{(1)}$ is a disconnected union of the curves D_i.

3.

Proof of the Theorem 1. Let L be a line in $\mathbb{C}^2$ transversally intersecting the curve D. Denote by $\gamma_i \subset L$ a circle of small radius with center at one of the common points of D_i and L. A cycle of $H_1(\mathbb{C}^2 \setminus D, \mathbb{Z})$ corresponding to γ_i will be denoted by the same letter. It is well known that the cycles $\gamma_1, \ldots, \gamma_k$ form a basis of $H_1(\mathbb{C}^2 \setminus D, \mathbb{Z})$. Similarly, for each i we choose a loop $\bar{\gamma}_i \subset \bar{F}_n^{-1}(L)$, which is a simple circuit on $\bar{F}_n^{-1}(L)$ around one of the common points of $\bar{F}_n^{-1}(L)$ and B_i.

Lemma 6. *The cycles $\bar{\gamma}_i$ are linearly independent in $H_1(X_n^0, \mathbb{Z})$.*

Proof. Let $F_* : H_1(X_n^0, \mathbb{Z}) \to H_1(\mathbb{C}^2 \setminus D, \mathbb{Z})$ be a homomorphism induced by the morphism F_n. Lemma 6 follows from that $F_*(\bar{\gamma}_i) = n_i \gamma_i$, where n_i is the multiplicity of ramification of $\bar{F}_n$ along B_i.

The proofs of Theorem 1 and Corollaries 1, 2 and 3 follow from Theorem 4 and from the following evident lemma.

Lemma 7. *Under the hypothesis of Theorem* 1,

i) *cycles $\bar{\gamma}_i$ belong to the kernel of a homomorphism*

$$i_* : H_1(X_n^0, \mathbb{Z}) \to H_1(X_n \setminus \mathrm{Sing}(X_n), \mathbb{Z}),$$

ii) *if for all j, $1 \le j \le k$, pre-images $F_n^{-1}(D_j)$ are irreducible, then the cycles $\bar{\gamma}_j$ generate the kernel of the homomorphism i_*.*

4.

In this section we will prove Theorem 2. For that we consider the exact sequences of chain complices [7]:

$$0 \longrightarrow C.(X_\infty) \xrightarrow{h^n - \mathrm{Id}} C.(X_\infty) \xrightarrow{F_{\infty,n*}} C.(X_n^0) \longrightarrow 0, \tag{4}$$

$$0 \longrightarrow C.(X_\infty) \xrightarrow{h - \mathrm{Id}} C.(X_\infty) \xrightarrow{F_{\infty*}} C.(\mathbb{C}^2 \setminus D) \longrightarrow 0 \tag{5}$$

and corresponding exact sequences of homology:

$$\to H_1(X_\infty) \xrightarrow{h^n - \mathrm{Id}} H_1(X_\infty) \longrightarrow H_1(X_n^0) \longrightarrow H_0(X_\infty) \longrightarrow 0, \quad (6)$$

$$\to H_1(X_\infty) \xrightarrow{h - \mathrm{Id}} H_1(X_\infty) \longrightarrow H_1(\mathbb{C}^2 \setminus D) \longrightarrow H_0(X_\infty) \longrightarrow 0, \quad (7)$$

where $F_\infty : X_\infty \to \mathbb{C}^2 \setminus D$ and $F_{\infty,n} : X_\infty \to X_n^0$ are infinite cyclic coverings determined by the epimorphism

$$\phi_* : \pi_1(\mathbb{C}^2 \setminus D) \to \pi_1(\mathbb{C}^1 \setminus \{0\}) \cong \mathbb{F}_1.$$

We have that $\dim H_0(X_\infty) = 1$, since X_∞ is connected, and therefore $\dim \ker(h - \mathrm{Id}) = k - 1$, if we take into account (7). It follows from (6) and (7) that

$$\dim H_1(X_n^0) = N(n, D) + k \quad (8)$$

for all n.

Now we will prove Theorem 2 in the case n is divisible by $n(D)$.

Lemma 8. *If $D \subset \mathbb{C}^2$ satisfies the hypothesis of Theorem 2, then, for all n divisible by $n(D)$, the dimension of the kernel of the epimorphism $i_{n*} : H_1(X_n^0, \mathbb{Q}) \to H_1(\tilde{X}_n, \mathbb{Q})$ is equal to k.*

Proof. It follows from Lemma 6 and Lemma 7 that the dimension of the kernel of the epimorphism $i_{n*} : H_1(X_n^0, \mathbb{Q}) \to H_1(\tilde{X}_n, \mathbb{Q})$ is equal to k in the case $n = n(D)$.

If n is divisible by $n(D)$, then

$$\dim H_1(X_n^0) = N(n(D), D) + k \quad (9)$$

and, consequently, this dimension does not depend on n.

On the other hand, for each n_2 dividing n_1, there exists a rational map $\tilde{F}_{n_1,n_2} : \tilde{X}_{n_1} \to \tilde{X}_{n_2}$, which can be assuming without loss of generality to be regular. The morphism $\tilde{F}_{n_1,n_2}$ induces a homomorphism $\tilde{F}_{n_1,n_2*} : H_1(\tilde{X}_{n_1}, \mathbb{Q}) \to H_1(\tilde{X}_{n_2}, \mathbb{Q})$, which is surjective. Indeed, since $\tilde{F}_{n_1,n_2}$ is surjective, then it is well known that

$$\tilde{F}^*_{n_1,n_2} : H^0(\tilde{X}_{n_2}, \Omega^1_{\tilde{X}_{n_2}}) \to H^0(X_{n_1}, \Omega^1_{X_{n_1}})$$

is an inclusion. Hence

$$\tilde{F}^*_{n_1,n_2} : H^1(\tilde{X}_{n_2}, \mathbb{Q}) \to H^1(\tilde{X}_{n_1}, \mathbb{Q})$$

is inclusion too. Thus we have the following inequality

$$\dim H_1(\tilde{X}_{n_1}) \geq \dim H_1(\tilde{X}_{n_2}) \quad (10)$$

for n_2 dividing n_1.

In any case it follows from Lemma 6 and Lemma 7 that $\dim \ker i_{n*} \geq k$. From (9) and (10) it follows that the dimension of the kernel of the epimorphism

$i_{n*} : H_1(X_n^0, \mathbb{Q}) \to H_1(\tilde{X}_n, \mathbb{Q})$ is equal to k in the case n is divisible by $n(D)$. Lemma 8 is proven.

We have the following commutative diagram

$$\begin{array}{ccc} H_1(X_\infty, \mathbb{C}) & \xrightarrow{\ h\ } & H_1(X_\infty, \mathbb{C}) \\ {\scriptstyle F_{\infty,n*}}\downarrow & & \downarrow{\scriptstyle F_{\infty,n*}} \\ H_1(X_n^0, \mathbb{C}) & \xrightarrow{\ h_{n*}\ } & H_1(X_n^0, \mathbb{C}) \end{array} \tag{11}$$

Consider the composition of $F_{\infty,n*}$ and i_*,

$$\begin{array}{ccc} H_1(X_\infty) & \xrightarrow{\ F_{\infty,n*}\ } & H_1(X_n^0) \\ & {\scriptstyle \tilde{F}_{\infty,n*}}\searrow & \downarrow{\scriptstyle i_*} \\ & & H_1(\tilde{X}_n) \end{array} \tag{12}$$

From Lemma 7 and Lemma 8 it follows that the cycles

$$\tilde{\gamma}_i = \sum_{j=1}^{n} h_{n*}^j \bar{\gamma}_i$$

generate $\ker i_*$, if n is divisible by $n(D)$, and are invariant under the action of h_{n*}, where the cycles $\bar{\gamma}_1, \dots, \bar{\gamma}_k$ were defined in section 3. Thus, from Lemma 8, and from (11) and (12) we obtain that if n is divisible by $n(D)$, then the set of eigen-values of h_{n*} acting on $H_1(\tilde{X}_n)$ coincides with $R(n(D), D)$.

Now let n be arbitrary. Put $n_1 = n \cdot n(D)$ and consider $H^1(\tilde{X}_n)$ instead of homology. The homomorphism

$$F_{n_1,n}^* : H^0(\tilde{X}_n, \Omega^1_{\tilde{X}_n}) \to H^0(\tilde{X}_{n_1}, \Omega^1_{\tilde{X}_{n_1}})$$

induced by the morphism $\tilde{F}_{n_1,n} : \tilde{X}_{n_1} \to \tilde{X}_n$ is an inclusion. Moreover, $\operatorname{Im} F_{n_1,n}^*$ coincides with the subspace of the invariant elements under the action of $h_{n_1}^n$, and, consequently, the same is true for the inclusion

$$F_{n_1,n}^* : H^1(\tilde{X}_n, \mathbb{Q}) \to H^1(\tilde{X}_{n_1}, \mathbb{Q}).$$

Thus we obtain that $\dim H^1(\tilde{X}_n, \mathbb{Q}) = N(n, D)$ for all n. Theorem 2 is proven.

References

[1] Deligne P., Théorie de Hodge II, III, Publ. Math. I. H. E. S. 40 (1972), 5–57; 44 (1974), 5–77.

[2] Esnault H., Fibre de Milnor d'un cone sur une courbe plane singuliére, Invent. Math. 54 (1982), 318–327.

[3] Kohno T., An algebraic computation of the Alexander polynomials of a plane algebraic curve, Proc. Japan Acad. Ser. A 59 (1983), 94–97.

[4] Kulikov Vic. S., The Alexander polynomials of plane algebraic curves, Izv. Ross. Akad. Nauk. Ser. Mat. 57 (1993), 76–101.

[5] Kulikov Vic. S., The Alexander polynomials of plane algebraic curves II, Preprint MPI/93-14.

[6] Libgober A., Alexander polynomials of plane algebraic curves and cyclic multiple planes, Duke Math. J. 49 (1982), 833–851.

[7] Milnor J., Infinite cyclic covers, Topology of manifolds, Prindle, Weber and Schmidt, Boston 1968, 115–133.

[8] Randell R., Milnor fibers and Alexander polynomials of plane curves, Proc. Symp. Pure Math. 40 (1983), 415–419.

[9] Zariski O., On the linear connection index of the algebraic surfaces $z^n = f(x, y)$, Proc. Nat. Acad. Sci. 15 (1929), 494–501.

On the arithmetic of singular $K3$-surfaces

I. R. Shafarevich

The aim of this paper is to discuss the influence of the field of definition of a $K3$-surface on its geometric properties. The geometric invariant we will consider here is the Neron-Severi (or Picard) group S_X of the surface X, consisting of divisor classes on the surface defined over the algebraic closure of the field of definition. By means of intersection numbers, the group S_X is converted to an integral lattice. We will dicuss here the following

Conjecture. For all $K3$-surfaces over a given algebraic number field K their Severi lattices S_X belong to a finite set of isomorphism classes.

This will be called *the weak form* of the conjecture. Its *strong form* asserts that the same is true if we consider $K3$-surfaces defined over all algebraic number fields of given degree $[K : Q] = n$.

The complexity of a $K3$-surface is measured by the rank ρ of its lattice S_X. It is known that always $1 \leq \rho \leq 20$, and the surfaces with $\rho = 20$ are called singular. We will prove:

Theorem 1. *The above conjecture is true in its strong form for singular $K3$-surfaces.*

We shall prove in our case even a stronger statement: there is only a finite number of isomorphism classes of singular $K3$-surfaces over the field of complex numbers $\mathbb{C}$ defined over an algebraic number field of given degree.

The proof consists of three steps: a) reduction of the theorem in the general case to the case of a singular Kummer $K3$-surface; b) reduction of the theorem for a singular Kummer $K3$-surface to an analogous statement for a singular abelian surface (i.e. abelian surface with $\rho = 4$); and c) the proof of the statement about singular abelian surfaces.

Step a). Let X be a singular $K3$-surface defined over an algebraic number field K of degree $[K : \mathbb{Q}] = n$. If some quantity N is bounded by a constant depending only on n we will designate it as: $N \ll \infty$. We prove that there always exists an extension K' of the field K with $[K' : K] \ll \infty$ over which all classes of the group S_X are defined. Indeed all classes of the group S_X are defined. Indeed all classes of the group S_X are defined over some finite extension $\tilde{K}$ of K since the group is finitely generated. We assume that $\tilde{K}/K$ is a Galois extension. Its Galois group G acts on S_X and defines a homomorphism $\varphi : G \to \mathrm{Aut}(S_X) \subset \mathrm{GL}(20, \mathbb{Z})$ with a finite image. According to a well-known observation (see e.g. [1]) a finite subgroup of $\mathrm{GL}(20, \mathbb{Z})$ has an order that is bounded, e.g. by $|\,\mathrm{GL}(20, \mathbb{Z}/3)|$. So the field K' that corresponds to $\mathrm{Ker}\,\varphi$ has the desired property. As $[K' : \mathbb{Q}] \ll \infty$ we can assume that all classes of the group S_X are already defined over K.

It was first observed by T. Shioda and H. Inose [2] that every singular $K3$-surface is either a Kummer surface or has a (singular) Kummer surface Y as a covering of degree 2. See also Nikulin [3] where our common work on this subject is contained as a part. Let $\varphi : Y \dashrightarrow X$ be this covering. We have to prove that Y and φ are defined over the field K', $[K' : Q] \ll \infty$ and that there can arise only a finite number of different isomorphism classes over $\mathbb{C}$ of surfaces X from a given surface Y. As is shown in [3], the ramification of the covering X consists of 8 disjoint smooth rational curves E_i, $i = 1, \dots, 8$. Let $\bar{K}(Y) = \bar{K}(X)(\sqrt{f})$. Then

$$(f) = \sum_{i=1}^{8} E_i + 2D. \tag{1}$$

We can assume that all classes in S_X defined over K. Since the E_i are unique in their classes, $\sigma E_i = E_i$ for all $\sigma \in \mathrm{Gal}(\bar{K}/K)$. On the other hand, $\sigma D \sim D$ and $\sigma D = D + (g_\sigma)$, $g_\sigma \in \bar{K}(X)^*$. From (1) we have that $(f^{\sigma-1}) = (g_\sigma^2)$ and as the g_σ are defined up to a constant factor, we can assume that $f^{\sigma-1} = g_\sigma^2$. For the cobord δg_σ of the cochain g_σ we then have $(\delta g_\sigma)^2 = 1$, $\delta g_\sigma \in (\mathrm{Br}K)_2$ where $\mathrm{Br}K$ is the Brauer group. But every element of order 2 in $\mathrm{Br}K$ is split by a quadratic extension of K and after such an extension we have $\delta g_\sigma = \delta \lambda_\sigma$, $\lambda_\sigma \in \bar{K}^*$. From Hilbert's "Satz 90" we have: $g_\sigma \lambda_\sigma^{-1} = h^{\sigma-1}$ with $h \in \bar{K}(X)^*$, $f^{\sigma-1} = \lambda_\sigma^2 h^{2(\sigma-1)}$, $\lambda_\sigma^2 = \mu^{\sigma-1}$ with $\mu \in \bar{K}^*$ and $f\mu^{-1}h^{-2} \in K(X)$. This proves that Y and φ are defined over an extension K, $[K' : \mathbb{Q}] \ll \infty$.

We have to prove now that only a finite number of different surfaces X (over $\mathbb{C}$) can have a given surface Y as a covering of degree 2. If i is the automorphism of the covering φ, then birationally $X = Y/i$. So we have to prove that there is only a finite number of different involutions up to conjugacy in $\mathrm{Aut}(Y)$. This follows from the fact that there exists a homomorphism $\mathrm{Aut}(Y) \to H$, $H \subset \mathrm{Aut}(S_Y)$ with finite kernel and cokernel and the group $\mathrm{Aut}(S_Y)$ is a semidirect product of the subgroup H and a certain normal subgroup W: $\mathrm{Aut}(S_Y) = H \cdot W$ ([4]). The group $\mathrm{Aut}(S_Y)$ is an arithmetic group and so our assertion is true for it (see [5]). From this it follows automatically for H and thus for $\mathrm{Aut}(Y)$.

Step b). This is completely parallel to step a), as for every Kummer surface $Y = \mathrm{Kum}(A)$ there exists by definition a covering $\varphi : A \dashrightarrow Y$ of degree 2 by an abelian surface A. The only difference is that the ramification of φ consists not of 8 but of 16 rational curves. Since Y was singular ($\rho = 20$), for the abelian surface A we have $\rho = 20 - 16 = 4$, i.e., A is also singular.

Step c). Let A be an abelian surface with $\rho = 4$ defined over an algebraic number field K of degree $[K : \mathbb{Q}] = n$. Then, over $\bar{K}$, A is isogenous to a product of two elliptic curves having complex multiplication in the same imaginary quadratic field L. We will consider a more general case of abelian variety of dimension g which over $\bar{K}$ is isogenous to a product of g elliptic curves with complex multiplication in the same imaginary quadratic field L. Then $\mathrm{End}_{\bar{K}}(A) \otimes \mathbb{Q} = M_g(L)$. We can apply to the group $\mathrm{End}_{\bar{K}}(A)$ the same argument as we applied in step a) to the group S_X and therefore assume that $\mathrm{End}_{\bar{K}}(A) = \mathrm{End}_K(A)$ (if necessary replacing K by an extension K' with $[K' : \mathbb{Q}] \ll \infty$). In our case $\mathrm{End}_K(A)$ is then an order in the full matrix algebra $M_g(L)$. In particular $\mathrm{End}_K(A)$ contains g elements of rank 1 whose sum is a scalar matrix. From this it follows that there exists an isogeny $\varphi : C_1 \times \cdots \times C_g \to A$ where φ and $C_1, \ldots, C_g$ are defined over K and $\mathrm{End}_K(C_i) = R_i$ are orders in the field L. We have to prove

α) that there are only a finite number of nonisomorphic such curves C_i over $\mathbb{C}$; and

β) that for given $C_1, \ldots, C_g$ there is, up to isomorphism, only a finite number of abelian varieties defined over a field K of given degree n which are isogenous $C_1 \times \cdots \times C_g$ so that the isogeny is defined over K.

Also in both cases the number is bounded by a constant depending only on n.

Statement α. An elliptic curve C is determined up to isomorphism over $\mathbb{C}$ by the value $j(C)$ of its absolute invariant j. According to the theory of complex multiplication $[L(j(C)) : L] = h(R)$, the class number of the order R where $R = \mathrm{End}(c)$, ([6]). As $j(C) \in K$, $h(R) \le n$. On the other hand $h(R) = h(L)\Psi(f)$, $\Psi(f) = f \prod_{p|f}(1 - (\frac{d}{p})p^{-1})$ where f is the conductor of the order R, d is the discriminant of the field L and the above formula is valid only for $d \ne -4, -3$, while in these exceptional cases the expression on the right-hand side should be divided by 2 or 3 ([7]). According to a theorem of Siegel [8], $\log h(L)/\log\sqrt{|d|} \to 1$ and so $|d|$ is bounded if $h(R) \le n$. It is quite elementary to verify that $\Psi(f) \to \infty$ when $f \to \infty$. So for $h(R) \le n$ there are only a finite number of orders R. For each order R there are exactly $h(R)$ values of $j(C)$. So all in all the number of nonisomorphic (over $\mathbb{C}$) elliptic curves with complex muliplication defined over some algebraic number field K with $[K : \mathbb{Q}] = n$ is $\ll \infty$.

Statement β. Let now the curves $C_1, \ldots, C_g$ be fixed. We are going to prove that for each isogeny $\varphi : C_1 \times \cdots \times C_g \to A$ defined over K there exists another isogeny $\varphi' : C_1 \times \cdots \times C_g \to A$ with $\deg \varphi' \ll \infty$. For this it is sufficient to prove that for each such isogeny there exists an endomorphism α of $C_1 \times \cdots \times C_g$ with $\mathrm{Ker}\,\alpha \supset \mathrm{Ker}\,\varphi$ and $(\mathrm{Ker}\,\alpha : \mathrm{Ker}\,\varphi) \ll \infty$.

We have to use and make explicit the fact that the isogeny φ is defined over K. This means that the group $\mathfrak{a} = \mathrm{Ker}\,\varphi$ is a G-invariant subgroup of the torsion subgroup of $(C_1 \times \cdots \times C_g)(\bar{K})$ where $G = \mathrm{Gal}(\bar{K}/K)$.

For this we use the principal theorem of the theory of complex multiplication as formulated in [9]. The action of G on the points of finite order of C_i defines the homomorphisms $G \to U(R_i = \prod_p (R_i \otimes \mathbb{Z}_p)^*$. Comparing with class field theory we obtain homomorphisms

$$\xi_i : C_K \to U(R_i)$$

where C_K is the idele class group, and corresponding homomorphisms

$$\varrho_i : I_K \to U(R_i)$$

where I_K is the idele group of K. ρ_i is trivial on the group of principal ideles P_K. The theory of complex multiplication asserts that the restriction of homomorphisms ρ_i to the group $U(K)$ of unitary ideles can be expressed as

$$\rho_i(x) = N_{K/L}(x)^{-1}\varepsilon_i(x) \tag{2}$$

where ε_i are homomorphisms of finite order: 2, 4 or 6 according to the number of roots of unity in L. From (2) it follows that a^m ($m = 2n$, $4n$ or $6n$) lies in the image of the Galois group G for every $a \in U(L)$. Therefore from the fact that the kernel $\mathfrak{a}$ of the isogeny φ is G-invariant it follows that $\mathfrak{a}$ is invariant under $U(L)^m$. However a subgroup invariant under a set X of operators is also invariant under the ring $\mathbb{Z}[X]$ generated by this set. We shall prove that the ring $\mathbb{Z}[U(L)^m]$ contains an order $R = R_f$ where f is the conductor of R and $f \ll \infty$. It is obvious that for every prime p, $(\mathbb{O}\otimes\mathbb{Z}_p : \mathbb{Z}[((\mathbb{O}\otimes\mathbb{Z}_p)^*)^m] \ll \infty$ where $\mathbb{O}$ is the ring of integers in L. It is necessary therefore to prove that this index is equal to 1 for almost all p. So we can assume that p is relatively prime to m and the discriminant of L. In this case it is sufficient to prove that $\mathbb{Z}[((\mathbb{F}_p \times \mathbb{F}_p)^*)^m] = \mathbb{F}_p \times \mathbb{F}_p$ and $\mathbb{Z}[(\mathbb{F}_{p^2}^*)^m] = \mathbb{F}_{p^2}$ for sufficiently large p. In the first case $((\mathbb{F}_p\times\mathbb{F}_p)^*)^m$ contains an element $(1, x)$, $x \neq 1$ if $(p-1)xm$. So our statement is false only for p with $(p-1)|m$, $p \leq m+1$. In the second case it is false only if $\mathbb{F}_{p^2}^m \subset \mathbb{F}_p$, $(p^2-1)|m(p-1)$, $p \leq m-1$. So one can choose R as an order defined by a fixed set of congruences, i.e. $R = Rf$, $f \ll \infty$.

We shall use the following formalism. Let T_e be the l-adic Tate module of $C_1 \times \cdots \times C_g$ and $T = \prod T_e$ the full Tate module. As usual let $V_e = T_e \otimes \mathbb{Q}_e$ and $V = \prod^* V_e$ where $\prod^*$ is the restricted product in the adelic sense. Then V/T is the torsion subgroup of $(C_1 \times \cdots \times C_g)(\bar{K})$. To each finite subgroup $\mathfrak{a} \subset V/T$ corresponds a subgroup A, $T \subset A \subset V$, $A/T = \mathfrak{a}$. Choose elements $e_1, \ldots, e_{2g} \in V$ in such a way that $\bigoplus_1^{2g} e_i\hat{\mathbb{Z}}$ is commensurable with T. Here $\hat{\mathbb{Z}} = \prod \mathbb{Z}_e$. To each subgroup $M \subset V$, commensurable with T, corresponds a free abelian group $\mathfrak{M} = M \bigcap (\oplus e_i\mathbb{Q})$ of rank $2g$. For two such groups M and N inclusions $M \subset N$ and $\mathfrak{M} \subset \mathfrak{N}$ (as subgroups of $\oplus e_i\mathbb{Q}$) are equivalent. Now choose the e_i in such a way that the group $\oplus e_i\mathbb{Z}$ is R-invariant. For instance put $\bigoplus_1^{2g} e_i\mathbb{Z} = \bigoplus_1^{g} f_iR$ where $f_i \in V(C_i)$ are such that $f_iR\otimes\hat{\mathbb{Z}}$ are commensurable with $T(C_i)$. Then a subgroup $M \subset V$, commensurable with T is R-invariant if and only if the corresponding subgroup $\mathfrak{M}$ is.

Therefore we can suppose that the subgroup $A \subset V$, $A/T = \mathfrak{a}$ and the corresponding subgroup $\mathfrak{A}$ are R-invariant. The vector space $\mathfrak{A} \otimes \mathbb{Q}$ is a module over $R \otimes \mathbb{Q} = L$ and therefore isomorphic to L^g. Among the R-submodules $\mathfrak{M} \subset L^g$

that are $\mathbb{Z}$-modules of rank $2g$ there is only a finite number of nonisomorphic submodules. (Theorem about the finiteness of the class number, see [10]). Let $\mathfrak{M}_1, \dots, \mathfrak{M}_h$ be their representatives. Here h and $\mathfrak{M}_1, \dots, \mathfrak{M}_h$ depend only on R, i.e. $C_1, \dots, C_g$ and n. We can choose them in such a way that corresponding submodules $M_i \subset V$ are contained in T. The module $\mathfrak{A}$ is isomorphic to one of them: $\mathfrak{A} \simeq \mathfrak{M}_i$. This isomorphism extends to an isomorphism $\mathfrak{A} \otimes \mathbb{Q} \simeq \mathfrak{M}_i \otimes \mathbb{Q} = L^g$ that is defined by an element α of $\mathrm{End}_L(L^g) = M_g(L) = \mathrm{End}(C_1 \times \cdots \times C_g) \otimes \mathbb{Q}$. Consequently $\alpha A = M_i$. As $A \supset T$, $\alpha A \supset \alpha T$ and as $\alpha A = M_i \subset T$ we see that $\alpha T \subset T$. From this it follows that $\alpha \in \mathrm{End}(C_1 \times \cdots \times C_g)$. On the other hand $\alpha A \subset T$ so $A \subset \alpha^{-1} T$ and $(\alpha^{-1}T : A) = (T : \alpha A) = (T : M_i)$. We see that $\mathrm{Ker}\, \alpha = \alpha^{-1}T/T \subset A/T = \mathfrak{a}$ and $(\mathrm{Ker}\, \alpha : \mathfrak{a}) \leq \mathrm{Max}(T : M_i) \ll \infty$. This completes the proof.

Remarks. 1. The stronger statement we proved for singular $K3$-surfaces: there is only a finite number of isomorphism classes of surfaces over $\mathbb{C}$ under the assumptions of the conjecture, can't be true in the general case. It is not true in the case $X = \mathrm{Kum}(A)$, $A = C_1 \times C_2$, where C_1 and C_2 are elliptic curves without complex multiplication defined over $\mathbb{Q}$.

2. After the end of the conference Ju. G. Zarhin showed me a shorter argument which in the case $g = 2$ avoids the proof of statement β). Namely according to T. Shioda and N. Mitani [11] every singular abelian surface is isomorphic over $\mathbb{C}$ to a product of elliptic curves: $A = C_1 \times C_2$. As in our case $\mathrm{End}_{\mathbb{C}}(A) = \mathrm{End}_K(A)$, the projections on C_1 and C_2 are defined over K. We have only to apply the statement α).

3. With every singular abelian surface (and with every singular $K3$-surface) there is associated an elliptic curve $C_A = H^2(A, \mathfrak{O})/j^* H^2(A, \mathbb{Z})$ where j^* corresponds to the inclusion $j : \mathbb{Z} \to \mathfrak{O}$. In [11] it is proved that in the decomposition $A = C_1 \times C_2$ one can assume $C_1 = C_A$. The preceding argument shows that C_A is defined over the field of definition of $\mathrm{End}\, A$. This increases the chances for its algebraic definition.

4. J.-P. Serre informed me that a question analogous to our conjecture was raised by Coleman concerning abelian varieties and their endomorpism rings. It is obvious that the proof of the statement β) gives a positive answer to this question.

References

[1] A. Speiser, Die Theorie der Gruppen von endlicher Ordnung, Berlin 1937.

[2] T. Shioda, H. Inose, On singular $K3$ surfaces, Complex Analysis and Algebraic Geometry, Cambridge University Press, 119–137.

[3] V. V. Nikulin, On rational maps between $K3$ surfaces, Constantin Caratheodory: An International Tribute, ed. by Th. M. Rassias, World Scientific Publ. 1991, 964–995.

[4] I. I. Piatetskij-Shapiro, I. R. Shafarevich, A Torelli theorem for algebraic surfaces of type $K3$, Izv. Akad. Nauk SSSR Ser. Mat. 35 (1971), 530–588.

[5] A. Borel, Arithmetic properties of linear algebraic groups, Proc. Internat. Congr. Mathematicians, Stockholm 1962, Uppsala 1993, 10–22.

[6] G. Shimura, Introduction to the arithmetic theory of automorphic functions, Princeton University Press 1971.

[7] R. Dedekind, P. G. Lejeune Dirichlet, Vorlesungen über Zahlentheorie, zweite Auflage, 1871, 163, 7.

[8] K. L. Siegel, Über die Classenzahl quadratischer Zahlkörper, Gesammelte Abhandlungen, Bd. 1, 406–409.

[9] J.-P. Serre, J. Tate, Good reduction of abelian varieties, Ann. of Math. 88 (1968), 492–517.

[10] C. W. Curtis, I. Reiner, Representation Theory of Finite Groups and Associative Algebas, Interscience Publishers 1962.

[11] T. Shioda, N. Mitani, Singular abelian surfaces and binary quadratic forms, in: Classification of Algebraic Varieties and Compact Complex Manifolds (ed. H. Popp), Lecture Notes in Math. 412, 259–287, Springer-Verlag, Berlin 1974.

The algebraic equation of type E_6 and Chebotarev density

Tetsuji Shioda

1. Introduction

Suppose we have an algebraic equation

$$F(X, \lambda) = X^n + a_1 X^{n-1} + \cdots + a_n = 0 \tag{1}$$

where $a_1, \ldots, a_n$ belong to a field $k_0 = \mathbb{Q}(\lambda)$ which is a purely transcendental extension of $\mathbb{Q}$ generated by some independent parameters λ. Let G be the Galois group of this algebraic equation, i.e. that of the splitting field $\mathcal{K}$ over k_0.

When we specialize the parameters λ to some rational numbers λ', (1) becomes an algebraic equation over $\mathbb{Q}$. By Hilbert's irreducibility theorem, its Galois group over $\mathbb{Q}$ is equal to G for most choices of λ' (see e.g. [4]). Take and fix such a λ', and let $K/\mathbb{Q}$ be the corresponding Galois extension.

According to the Chebotarev density theorem, the density of prime numbers p such that the Frobenius transformation ϕ_p of K at p belongs to a given conjugacy class $C \subset G$ is equal to $|C|/|G|$. (Here $|A|$ denotes the cardinality of a set A). In particular, the prime numbers p for which ϕ_p is the identity (such a p is called *completely split*) has the density $1/|G|$.

Thus it might be of some interest to ask if one can arrange $K/\mathbb{Q}$ in the above situation in such a way that there is a small p, compared with the size of $|G|$, which is completely split.

In this note, we give an example of this sort for the algebraic equation of of degree 27 of type E_6, where $G = W(E_6)$ is the Weyl group of type E_6 with order $|G| = 51840$. This equation arises naturally from the Mordell–Weil lattices of type E_6, which can be also interpreted as the algebraic equation of 27 lines on a smooth cubic surface.

A numerical example of Galois extension of $\mathbb{Q}$ with Galois group $W(E_6)$ has been constructed in our previous paper [6]. In the present note, we construct a new example where a completely split p occurs as early as $p = 13$. The idea is to combine the construction for (a) *big Galois* and (b) *small Galois* discussed in [6, Part II].

2. Algebraic equations of type E_6

The polynomial $F(X)$ of degree 27 below is what we called the *universal polynomial of type* E_6 in [5]. The coefficients belong to the ring $R = \mathbb{Z}[p_0, p_1, p_2, q_0, q_1, q_2]$. It arises naturally from the Mordell–Weil lattice of the elliptic curve E_λ over $k(t)$ defined by the equation:

$$y^2 = x^3 + (p_0 + p_1 t + p_2 t^2)x + (q_0 + q_1 t + q_2 t^2 + t^4) \tag{2}$$

where we set $\lambda = (p_0, p_1, p_2, q_0, q_1, q_2)$, $k_0 = \mathbb{Q}(\lambda) = \mathbb{Q}(p_0, p_1, p_2, q_0, q_1, q_2)$, and k denotes the algebraic closure of k_0.

For general λ, the Mordell–Weil lattice $E_\lambda(k(t))$ is isomorphic to the dual lattice E_6^* of the root lattice E_6. As a consequence, there are exactly 27 rational points $P = (x, y) \in E_\lambda(k(t))$ of the form:

$$x = at + b, \quad y = t^2 + dt + e \quad (a, b, d, e \in k). \tag{3}$$

The algebraic equation $F(X) = 0$ is obtained as follows. First, substituting (3) into (2) and comparing the coefficients of t^i, we get several relations among a, b, d, e. Then elimination of d, e, b from them yields the desired equation $F(a) = 0$. In other words, the 27 roots of the polynomial $F(X) = F(X, \lambda)$ below are precisely the t-coefficient a of the x-coordinate of such P. We know that a is the most important parameter on which other b, d, e depend rationally (cf. also [8]).

By the coordinate change $y' = y - t^2$, $(x : y' : t : 1) = (X, Y, Z, W)$, the equation (2) is transformed to the equation of a cubic surface V_λ:

$$Y^2W + 2YZ^2 = X^3 + (p_0W^2 + p_1ZW + p_2Z^2)X + (q_0W^2 + q_1ZW + q_2Z^2)W \tag{4}$$

The 27 lines on V_λ are then given by the equations

$$X = aZ + bW, \quad Y = dZ + eW. \tag{5}$$

Therefore the algebraic equation $F(X) = 0$ becomes that of 27 lines on the cubic surface (4) which contains 6 essential parameters p_i, q_j. (For a historical account of the algebraic equation of 27 lines on a cubic surface, we refer to [1], [2] and [3].)

$$
\begin{aligned}
F(X) = {} & \\
& X^{27} \\
& + X^{25}(12p_2) \\
& + X^{23}(60{p_2}^2) \\
& + X^{22}(-48p_1) \\
& + X^{21}\left(96q_2 + 168{p_2}^3\right) \\
& + X^{20}(-336p_1p_2) \\
& + X^{19}\left(480p_0 + 294{p_2}^4 + 528p_2q_2\right) \\
& + X^{18}\left(-1344q_1 - 1008p_1{p_2}^2\right) \\
& + X^{17}\left(144{p_1}^2 + 2304p_0p_2 + 336{p_2}^5 + 1152{p_2}^2q_2\right) \\
& + X^{16}\left(-1680p_1{p_2}^3 - 5568p_2q_1 - 768p_1q_2\right) \\
& + X^{15}\left(17280q_0 + 608{p_1}^2p_2 + 4768p_0{p_2}^2 + 252{p_2}^6 + 1200{p_2}^3q_2 - 1248{q_2}^2\right) \\
& + X^{14}\left(2304p_0p_1 - 1680p_1{p_2}^4 - 8832{p_2}^2q_1 - 2688p_1p_2q_2\right)
\end{aligned}
$$

$$
\begin{aligned}
&+ X^{13}\left(976p_1{}^2p_2{}^2 + 5696p_0p_2{}^3 + 120p_2{}^7 + 43776p_2q_0 + 3264p_1q_1 + 12288p_0q_2\right.\\
&\quad\left. + 480p_2{}^4q_2 - 4800p_2q_2{}^2\right)\\
&+ X^{12}\left(832p_1{}^3 + 5888p_0p_1p_2 - 1008p_1p_2{}^5 - 6528p_2{}^3q_1 - 3072p_1p_2{}^2q_2\right.\\
&\quad\left. + 10752q_1q_2\right)\\
&+ X^{11}\left(-34560p_0{}^2 + 704p_1{}^2p_2{}^3 + 4384p_0p_2{}^4 + 33p_2{}^8 + 41472p_2{}^2q_0\right.\\
&\quad\left. + 2688p_1p_2q_1 + 4224p_1{}^2q_2 + 34560p_0p_2q_2 - 144p_2{}^5q_2 - 6720p_2{}^2q_2{}^2\right)\\
&+ X^{10}\left(2560p_1{}^3p_2 + 3584p_0p_1p_2{}^2 - 336p_1p_2{}^6 + 64512p_1q_0 - 70656p_0q_1\right.\\
&\quad\left. - 2112p_2{}^4q_1 - 768p_1p_2{}^3q_2 + 23040p_2q_1q_2 + 8448p_1q_2{}^2\right)\\
&+ X^9\left(-18944p_0p_1{}^2 - 39936p_0{}^2p_2 + 176p_1{}^2p_2{}^4 + 2176p_0p_2{}^5 + 4p_2{}^9\right.\\
&\quad + 22528p_2{}^3q_0 - 5504p_1p_2{}^2q_1 - 40704q_1{}^2 + 8960p_1{}^2p_2q_2 + 32768p_0p_2{}^2q_2\\
&\quad\left. - 192p_2{}^6q_2 + 110592q_0q_2 - 3840p_2{}^3q_2{}^2 + 5120q_2{}^3\right)\\
&+ X^8\left(2688p_1{}^3p_2{}^2 - 1536p_0p_1p_2{}^3 - 48p_1p_2{}^7 + 82944p_1p_2q_0 + 4608p_1{}^2q_1\right.\\
&\quad - 64512p_0p_2q_1 - 192p_2{}^5q_1 - 73728p_0p_1q_2 + 768p_1p_2{}^4q_2 + 13824p_2{}^2q_1q_2\\
&\quad\left. + 16128p_1p_2q_2{}^2\right)\\
&+ X^7\left(-2560p_1{}^4 - 16384p_0p_1{}^2p_2 - 20992p_0{}^2p_2{}^2 - 32p_1{}^2p_2{}^5 + 608p_0p_2{}^6\right.\\
&\quad - 165888p_0q_0 + 9600p_2{}^4q_0 - 6144p_1p_2{}^3q_1 - 38400p_2q_1{}^2 + 5376p_1{}^2p_2{}^2q_2\\
&\quad + 10752p_0p_2{}^3q_2 - 48p_2{}^7q_2 + 156672p_2q_0q_2 - 15360p_1q_1q_2 - 56832p_0q_2{}^2\\
&\quad\left. - 480p_2{}^4q_2{}^2 + 9984p_2q_2{}^3\right)\\
&+ X^6\left(49152p_0{}^2p_1 + 1024p_1{}^3p_2{}^3 - 1792p_0p_1p_2{}^4 + 21504p_1p_2{}^2q_0\right.\\
&\quad + 10240p_1{}^2p_2q_1 - 19456p_0p_2{}^2q_1 - 110592q_0q_1 - 10240p_1{}^3q_2\\
&\quad\left. - 57344p_0p_1p_2q_2 + 384p_1p_2{}^5q_2 + 1536p_2{}^3q_1q_2 + 6912p_1p_2{}^2q_2{}^2 - 21504q_1q_2{}^2\right)\\
&+ X^5\left(-1536p_1{}^4p_2 - 4608p_0p_1{}^2p_2{}^2 - 3072p_0{}^2p_2{}^3 - 16p_1{}^2p_2{}^6 + 64p_0p_2{}^7\right.\\
&\quad + 27648p_1{}^2q_0 - 110592p_0p_2q_0 + 2304p_2{}^5q_0 - 9216p_0p_1q_1 - 1344p_1p_2{}^4q_1\\
&\quad - 12288p_2{}^2q_1{}^2 + 73728p_0{}^2q_2 + 768p_1{}^2p_2{}^3q_2 + 55296p_2{}^2q_0q_2\\
&\quad\left. + 10752p_1p_2q_1q_2 - 19200p_1{}^2q_2{}^2 - 46080p_0p_2q_2{}^2 + 192p_2{}^5q_2{}^2 + 4608p_2{}^2q_2{}^3\right)\\
&+ X^4\left(8192p_0p_1{}^3 + 16384p_0{}^2p_1p_2 + 64p_1{}^3p_2{}^4 - 256p_0p_1p_2{}^5 + 3072p_1p_2{}^3q_0\right.\\
&\quad - 512p_1{}^2p_2{}^2q_1 - 1024p_0p_2{}^3q_1 - 36864p_2q_0q_1 - 39936p_1q_1{}^2 - 4096p_1{}^3p_2q_2\\
&\quad - 8192p_0p_1p_2{}^2q_2 + 73728p_1q_0q_2 + 24576p_0q_1q_2 - 768p_1p_2{}^3q_2{}^2 - 3072p_2q_1q_2{}^2\\
&\quad\left. - 18432p_1q_2{}^3\right)\\
&+ X^3\left(-16384p_0{}^3 + 256p_1{}^4p_2{}^2 - 1024p_0p_1{}^2p_2{}^3 + 256p_0{}^2p_2{}^4\right.\\
&\quad - 6144p_1{}^2p_2q_0 - 18432p_0p_2{}^2q_0 + 256p_2{}^6q_0 - 110592q_0{}^2 + 15360p_1{}^3q_1\\
&\quad - 14336p_0p_1p_2q_1 - 128p_1p_2{}^5q_1 - 2560p_2{}^3q_1{}^2 + 8192p_0p_1{}^2q_2 + 24576p_0{}^2p_2q_2\\
&\quad + 128p_1{}^2p_2{}^4q_2 - 256p_0p_2{}^5q_2 + 9216p_2{}^3q_0q_2 + 2048p_1p_2{}^2q_1q_2 - 30720q_1{}^2q_2\\
&\quad\left. - 2560p_1{}^2p_2q_2{}^2 - 7680p_0p_2{}^2q_2{}^2 + 55296q_0q_2{}^2 - 256p_2{}^3q_2{}^3 - 6912q_2{}^4\right)\\
&+ X^2\left(-1024p_1{}^5 + 4096p_0p_1{}^3p_2 - 12288p_1p_2q_1{}^2 + 24576p_1{}^2q_1q_2\right)\\
&+ X\left(-4096p_0{}^2p_1{}^2 - 3072p_1{}^2p_2{}^2q_0 + 2048p_1{}^3p_2q_1 - 1024p_0p_1p_2{}^2q_1\right.\\
&\quad - 36864p_1q_0q_1 + 12288p_0q_1{}^2 - 256p_2{}^4q_1{}^2 - 2048p_1{}^4q_2 + 4096p_0p_1{}^2p_2q_2\\
&\quad\left. + 512p_1p_2{}^3q_1q_2 - 6144p_2q_1{}^2q_2 - 256p_1{}^2p_2{}^2q_2{}^2 + 9216p_1q_1q_2{}^2\right)\\
&+ 1024\left(4p_1{}^3q_0 - 4p_0p_1{}^2q_1 - p_1p_2{}^2q_1{}^2 - 4q_1{}^3 + 2p_1{}^2p_2q_1q_2 - p_1{}^3q_2{}^2\right).
\end{aligned}
$$

Let $\mathcal{K}_\lambda$ denote the splitting field of $F(X) = F(X, \lambda)$ over $k_0 = \mathbb{Q}(\lambda) = \mathbb{Q}(p_i, q_j)$. The fundamental theorems for the algebraic equation of type E_6 ([5], Th. 10.3, 10.4) state that, for λ generic, $\mathcal{K}_\lambda$ is a Galois extension of k_0 with the Galois group $W(E_6)$, the Weyl group of type E_6; the latter is a finite reflection group of order

51840 and has a simple subgroup of index 2. Moreover, if $P_1, \dots, P_6$ are 6 rational points of the form (3) which are linearly independent in the Mordell–Weil lattice, then the corresponding $a_1, \dots, a_6$ are algebraically independent and generate $\mathcal{K}_\lambda$ over $\mathbb{Q}$, i.e. $\mathcal{K}_\lambda = \mathbb{Q}(a_1, \dots, a_6)$. Furthermore p_i, q_j are polynomials of $a_1, \dots, a_6$ with rational coefficients which form a set of the fundamental invariants of $W(E_6)$ with weights $wt(p_i) = 8 - 3i$, $wt(q_i) = 12 - 3i (i = 0, 1, 2)$.

The fundamental theorems have the standard applications (a) and (b) below, in perfect analogy to the familiar case of generic algebraic equation of degree n (corresponding to type A_{n-1}).

(a) *big Galois*: By specializing the parameters p_i, q_j to suitable rational numbers, we get a Galois extension of $\mathbb{Q}$ with Galois group $W(E_6)$. By Hilbert irreducibility theorem, this is the case for most choice of specialization. Moreover every extension of $\mathbb{Q}$ with Galois group $W(E_6)$ is obtained this way (see [6]).

(b) *small Galois*: By specializing $a_1, \dots, a_6$ to rational numbers in such a way that the Mordell–Weil lattice does not degenerate, we get the trivial extension of $\mathbb{Q}$, which means that the Mordell–Weil group $E(\mathbb{Q}(t))$ has rank 6 (without making any constant field extension) (see [5]).

3. Construction method of examples

It follows from (a) that for most choice of $p_i, q_j \in \mathbb{Q}$, we get a Galois extension of $\mathbb{Q}$ with Galois group $W(E_6)$. As in [6], there is a criterion to verify this for given p_i, q_j by considering $F(X)$ mod p for various primes p (see below). But this does not guarantee if the trivial Frobenius transformation $\phi_p = 1$ occurs for a small p. So we start from (b).

Step 1. Choosing $a_1, \dots, a_6 \in \mathbb{Q}$ as in (b), determine $\lambda' = (p'_i, q'_j) \in \mathbb{Q}^6$ by the explicit formula of the fundamental invariants and let $E' = E_{\lambda'}$. Then $E'(\mathbb{Q}(t)) \simeq E_6^*$ has rank 6 and has 27 rational points of the form (3) with rational coefficients. The algebraic equation $F(X, \lambda') = 0$ splits into 27 linear factors over $\mathbb{Q}$.

Step 2. Consider a non-degenerate reduction of E' modulo a suitable prime p (in the sense of Mordell–Weil lattices). Namely, reducing E' modulo p, we get an elliptic curve $\bar{E}' = E'_p$ over $\mathbb{F}_p(t)$ such that the Mordell–Weil group $\bar{E}'(\mathbb{F}_p(t))$ has again rank 6. In fact, this is the case for almost all p. Fix one such p.

Step 3. Next lift $\bar{E}'$ back to characteristic 0. Namely consider $E = E_\lambda$ with $\lambda = (p_i, q_j) \in \mathbb{Q}^6$ such that $p_i \equiv p'_i, q_j \equiv q'_j \bmod p$. By (a), we can expect that for a suitable choice of λ, the algebraic equation $F(X, \lambda) = 0$ over $\mathbb{Q}$ has the Galois group $W(E_6)$. In that case, the chosen p is clearly a completely split prime in the corresponding extension, since E and E' have the same reduction $\bar{E}'$ mod p.

4. Examples

Let us write down some explicit examples by the above method.

Example 1. Take $p_0 = 13$, $p_1 = p_2 = q_2 = 0$, $q_0 = 8$, $q_1 = 13$. The elliptic curve E over $\mathbb{Q}(t)$ is given by

$$E : y^2 = x^3 + 13x + 8 + 13t + t^4, \tag{6}$$

and we have

$$\begin{aligned} F(X) = {} & X^{27} + 6240X^{19} - 17472X^{18} + 138240X^{15} - 5840640X^{11} \\ & - 11940864X^{10} - 6878976X^9 - 17252352X^7 - 11501568X^6 \\ & - 43073536X^3 + 26996736X - 8998912. \end{aligned} \tag{7}$$

The splitting field K of this equation $F(X) = 0$ over $\mathbb{Q}$ has the Galois group $W(E_6)$, and $p = 13$ is completely split in K.

This can be verified as follows. As the auxiliary curve E' in Step 1, we take

$$E' : y^2 = x^3 + (-59475 - 78t^2)x + 2848750 + 18226t^2 + t^4 \tag{8}$$

which has been studied in detail in [8, §8]. The corresponding cubic surface V' is given by the equation:

$$Y^2W + 2YZ^2 = X^3 + X(-59475W^2 - 78Z^2) + 2848750W^3 + 18226Z^2W, \tag{9}$$

and all the 27 lines are defined over $\mathbb{Q}$. This is constructed from the data $a_i = 2(9-i)$ $(1 \le i \le 6)$ such that the 6 lines L_i on V' corresponding to a_i are mutually disjoint. The 27 roots of $F(X, \lambda') = 0$ are as follows:

$$0^3, \pm 2^2, \pm 4^2, \pm 6^2, \pm 8^2, \pm 10, \pm 12, \pm 14, \pm 16, \tag{10}$$

where for instance $\pm 4^2$ means that each of ± 4 appears with multiplicity 2.

Reducing (8) mod $p = 13$, we have

$$\bar{E}' : y^2 = x^3 + 8 + t^4 \tag{11}$$

over $\mathbb{F}_p(t)$, which is obviously non-degenerate and of type E_6.

The equation (6) is a lifting of (11), so $p = 13$ is completely split (note that $F(X)$ and $F(X, \lambda')$ are equal mod $p = 13$).

It remains to show that $F(X) = 0$ has the full $W(E_6)$ as the Galois group. For this, we look at $F_\ell = F(X)$ mod ℓ for various prime number ℓ. We find that for $\ell = 5$ (resp. 51), F_ℓ is decomposed into 3 irreducible factors of degree 9 (resp. 5 irreducible factors of degrees 2,5,5,5,10). This information suffices to conclude that the Galois group is equal to $W(E_6)$ by [6, Lem. 7.5].

Taking the same E' in Step 1 as above and choosing the prime number $p = 17$ in Step 2, we have the reduced curve

$$\bar{E}' : y^2 = x^3 + (8 + 7t^2)x + 9 + 2t^2 + t^4. \tag{12}$$

When we lift it back to characteristic 0, this equation (12) itself viewed over $\mathbb{Q}(t)$ is not appropriate for our purpose, since $p_1 = q_1 = 0$ implies $F(X) = X^3 \cdot (X^{24} + \cdots)$

which is reducible and cannot have the big Galois group $W(E_6)$. So modifying a little, we try (13) below, and we obtain a second example:

Example 2. Consider the elliptic curve E over $\mathbb{Q}(t)$ given by

$$E : y^2 = x^3 + (8 + 7t^2)x + 9 + 17t + 2t^2 + t^4. \tag{13}$$

Then we have

$$\begin{aligned} F(X) \; = \; & X^{27} + 84X^{25} + 2940X^{23} + 57816X^{21} + 717126X^{19} - 22848X^{18} \\ & + 5889072X^{17} - 662592X^{16} + 32490332X^{15} - 7357056X^{14} \\ & + 119580040X^{13} - 37699200X^{12} + 288236801X^{11} - 90331200X^{10} \\ & + 471155868X^{9} - 93242688X^{8} + 618051168X^{7} - 130124800X^{6} \\ & + 535579136X^{5} - 82026496X^{4} - 44281088X^{3} - 174084352X \\ & - 20123648. \end{aligned} \tag{14}$$

By the same argument as above, this equation is shown to have the Galois group $W(E_6)$, for which $p = 17$ is a completely split prime.

This example is mentioned for the following reason. Recall that there are 25 conjugacy classes in the group $W(E_6)$ (cf. [3], [9]). In Example 2, not only the identity class occurs for a small p, but also 23 (out of 25) conjugacy classes occur as the classes containing the Frobenius transformations ϕ_p already for $p < 1000$. The 2 missing classes are No.3 and 11 in the notation of [3] with density 1/648 or 1/1152. The verification can be done by the same method as in [7, §4].

References

[1] Henderson, A., The twenty-seven lines upon the cubic surface, Cambridge Univ. Press 1911.

[2] Jordan, C., Traité des substitutions et des équations algébriques, Gauthier-Villars, Paris 1870, 1957.

[3] Manin, Ju., Cubic Forms, North-Holland 1974.

[4] Serre, J.-P., Lectures on the Mordell–Weil Theorem, Vieweg 1989.

[5] Shioda, T., Construction of elliptic curves with high rank via the invariants of the Weyl groups, J. Math. Soc. Japan 43 (1991), 673–719.

[6] —, Theory of Mordell–Weil lattices, Proc. ICM Kyoto-1990, Springer 1991, vol. I, 473–489.

[7] —, Some remarks on elliptic curves over function fields, Proc. Journée Arithmetique 91, Genève, Astérisque 209 (1992), 99–114.

[8] —, Weierstrass transformations and cubic surfaces, Preprint 1994.

[9] Swinnerton-Dyer, H. P. F., The zeta function of a cubic surface over a finite field, Proc. Cambridge Philos. Soc. 63 (1967), 55–71.

Independent systems of derivations and Lie algebra representations

*Serge Skryabin**

Introduction

A unifying approach to various classes of Cartan type Lie algebras has evolved in recent years (see [22]). It seems to be quite a general principle that in the study of a question related to Cartan type algebras there is some essential part that applies to the class of Cartan type algebras as a whole regardless of any specific features exhibited in each particular case. From this viewpoint it should be of no surprise that the earlier results on representations of Cartan type Lie algebras of formal vector fields due to A. N. Rudakov and those on representations of finite dimensional Cartan type Lie algebras of positive characteristic due to several authors are fragments of a general theory.

In this paper we study representations of a Lie algebra W of derivations of a commutative algebra R over a field k such that W is a module over R. Our results are more advanced in case when $W = R\mathfrak{D}$ is generated as a module over R by a finite system of pairwise commuting independent derivations. In this case there is a certain class of modules whose representation theory is, roughly speaking, independent of a particular choice of R. In Section 1 we give an explicit construction of modules and introduce a certain category $\mathcal{C}$ which provides enough flexibility to work. In Section 2 we show how the modules we have constructed are related to the coinduced and induced modules. The notion of independence of operators is discussed in Section 3. Although simple in nature, it provides a very efficient tool to handle the problems we are dealing with. The main results on description of submodules and homomorphisms are presented in Section 4. They have been obtained as an extension of the work announced earlier [20]. In particular, the irreducibility of the representations of the Lie algebra of all formal vector fields induced and coinduced from the irreducible representations of its standard maximal subalgebra is an easy consequence of our results. In Section 5

* The author acknowledges a partial support of International Science Foundation (through grant NNPOOO) and Novosibirsk Grant Centre

we consider applications to the representations of generalized Witt algebras $W(\mathbf{m})$ in positive characteristic.

In Section 6 we make an attempt to develop an approach to representations without previous restriction on W. We have to narrow the class of representations. However, those that we study are the most important ones because they correspond to the representations of the Lie algebras of vector fields on manifolds in the modules of sections of vector bundles. Section 7 is devoted to the exceptional modules, which are the modules of differential forms. In particular, we generalize the classification of unary invariant differential operators.

The author believes that the approach developed here provides a natural framework to study the other classical problems such as the classification of binary invariant operators etc. Our methods are quite workable for the Lie algebras of special, hamiltonian and contact types as well.

1. An explicit construction of modules

Let R be a commutative, associative and unital algebra over a commutative ring k, and $\partial_1, \ldots, \partial_n$ pairwise commuting derivations of R. Let $\mathfrak{D}$ be a free k-module of rank n with a fixed basis, the elements of which we identify with $\partial_1, \ldots, \partial_n$. We regard $\mathfrak{D}$ as a commutative Lie k-algebra operating on R as a Lie algebra of derivations. Define $R\mathfrak{D}$ to be the Lie algebra over k with the underlying k-module $R \otimes_k \mathfrak{D}$ and the bracket operation given by the rule

$$[f \otimes \partial_i,\, g \otimes \partial_j] = f\partial_i(g) \otimes \partial_j - g\partial_j(f) \otimes \partial_i\,, \qquad f,\, g \in R.$$

Assigning to $f \otimes \partial_i$ the derivation $f\partial_i$ of R, we get a Lie algebra homomorphism $R\mathfrak{D} \to \operatorname{Der} R$. In case that $\partial_1, \ldots, \partial_n$ are linearly independent over R we thus may identify $R\mathfrak{D}$ with a Lie subalgebra of $\operatorname{Der} R$ generated as a module over R by $\partial_1, \ldots, \partial_n$.

The reason for dissociating $\partial_1, \ldots, \partial_n$ from their specific meaning of being derivations of R is that we shall consider a representation of $R\mathfrak{D}$ in a different so-called completed divided power algebra $\widehat{O}_n(k)$. All parallel constructions then apply, allowing us to work on a formal level. The divided power algebra $O_n(k)$ in n indeterminates $x_1, \ldots, x_n$ is a commutative associative k-algebra with a unity which can be defined by means of generators $x_i^{(r)}$ ($1 \le i \le n$, r a nonnegative integer) and a system of relations

$$x_i^{(0)} = 1, \qquad x_i^{(r)} x_i^{(s)} = \binom{r+s}{r} x_i^{(r+s)} \qquad (1 \le i \le n, \quad r, s \ge 0).$$

The monomials in x's of degree at least l span an ideal $\mathfrak{m}^{(l)}$. The algebra $\widehat{O}_n(k)$ is the completion of $O_n(k)$ with respect to the topology defined by those ideals. In case k is an algebra over a field of characteristic zero the assignment $x_i^{(r)} \mapsto x_i^r/r!$ yields isomorphisms of $O_n(k)$ and $\widehat{O}_n(k)$ onto the polynomial and the formal power series algebras in $x_1, \ldots, x_n$ respectively. We associate with each ∂_i a derivation of $O_n(k)$ and by continuity a derivation of $\widehat{O}_n(k)$ putting

$$\partial_i 1 = 0, \qquad \partial_i x_j^{(r)} = \delta_{ij} x_j^{(r-1)} \qquad (1 \le i, j \le n, \quad r > 0).$$

Then $W_n(k) = O_n(k)\mathfrak{D}$ and $\widehat{W}_n(k) = \widehat{O}_n(k)\mathfrak{D}$ are called the Lie algebras of special derivations of $O_n(k)$ and $\widehat{O}_n(k)$ respectively [15].

Denote by $A(n)$ the set of all n-tuples of nonnegative integers. We use componentwise operations on $A(n)$. So, given $\alpha = (\alpha_1, \ldots, \alpha_n)$ and $\beta = (\beta_1, \ldots, \beta_n)$ in $A(n)$, we put

$$\alpha + \beta = (\alpha_1 + \beta_1, \ldots, \alpha_n + \beta_n), \qquad \alpha - \beta = (\alpha_1 - \beta_1, \ldots, \alpha_n - \beta_n),$$

$$|\alpha| = \sum_{i=1}^{n} \alpha_i, \qquad \alpha! = \prod_{i=1}^{n} \alpha_i!, \qquad \binom{\alpha}{\beta} = \frac{\alpha!}{\beta!(\alpha-\beta)!} = \prod_{i=1}^{n} \binom{\alpha_i}{\beta_i}.$$

We write $\alpha \le \beta$ if $\alpha_i \le \beta_i$ for all i and $\alpha < \beta$ if in addition $\alpha_i < \beta_i$ for at least one i. Denote by ϵ_i the n-tuple whose i-th component is equal to 1, the others are zero. Denote also

$$\partial^{\alpha} = \prod_{i=1}^{n} \partial_i^{\alpha_i}, \qquad x^{(\alpha)} = \prod_{i=1}^{n} x_i^{(\alpha_i)}.$$

The monomials ∂^{α} (respectively $x^{(\alpha)}$), $\alpha \in A(n)$, constitute a basis for the universal enveloping algebra $U(\mathfrak{D})$ (respectively the algebra $O_n(k)$) as a k-module. The elements of $\widehat{O}_n(k)$ are infinite formal linear combinations $\sum \lambda_\alpha x^{(\alpha)}$ with $\lambda_\alpha \in k$. The monomials $x^{(\alpha)}$ with $|\alpha| \ge l$ constitute a basis for $\mathfrak{m}^{(l)}$, and so the closure $\widehat{\mathfrak{m}}^{(l)}$ of $\mathfrak{m}^{(l)}$ in $\widehat{O}_n(k)$ consists of those infinite linear combinations having zero coefficients λ_α for all α with $|\alpha| < l$. The subalgebras $\widehat{W}_n(k)_l = \widehat{\mathfrak{m}}^{(l+1)}\widehat{W}_n(k)$, $l \ge -1$, form a standard filtration in $\widehat{W}_n(k)$.

Let σ_V be a representation of $\widehat{W}_n(k)_0$ in a k-module V. We call the $\widehat{W}_n(k)_0$-module V discrete if for any $v \in V$ there exists $l \ge 0$ such that $\sigma_V(\widehat{W}_n(k)_l)v = 0$. If V is an irreducible discrete $\widehat{W}_n(k)_0$-module then there exists $l \ge 0$ such that $\sigma_V(\widehat{W}_n(k)_l) = 0$. The least possible l with the just stated property is called then the height of V [19]. Assuming V to be discrete, we define an endomorphism $\rho(D)$ of $R \otimes_k V$ for each $D \in R\mathfrak{D}$ by the rule

$$\rho(f \otimes \partial_i) = f\partial_i \otimes \mathrm{id}_V + \sum_{\alpha > 0} \partial^{\alpha} f \otimes \sigma_V(x^{(\alpha)}\partial_i), \qquad f \in R.$$

Here we consider $\varphi \otimes \psi$ as an endomorphism of $R \otimes_k V$ in a natural way for any k-linear endomorphism φ of R and k-linear endomorphism ψ of V, and we identify the elements of R with the corresponding multiplication operators. The infinite sum on the right makes sense because for any $v \in V$ we have $\sigma_V(x^{(\alpha)}\partial_i)v = 0$ for all but a finite number of α's.

It is a matter of straightforward computations to verify that ρ is a representation of the Lie algebra $R\mathfrak{D}$. The deliberations in Section 2 will make this fact quite transparent, however. Denote by $\mathcal{R}V$ the $R\mathfrak{D}$-module $(R \otimes_k V, \rho)$.

The description of modules is especially simple in case when $\widehat{W}_n(k)_1$ annihilates V. Such modules V correspond to the modules over the factor algebra

$\widehat{W}_n(k)_0/\widehat{W}_n(k)_1$, which is identified with the full Lie algebra of linear transformations of the k-module $\widehat{W}_n(k)/\widehat{W}_n(k)_0 \cong \mathfrak{D}$ by means of the representation induced by the adjoint one. Denoting by E_{ij} the linear transformation of $\mathfrak{D}$ such that $E_{ij}\partial_{j'} = \delta_{jj'}\partial_i$, we see that $x_j\partial_i$ operates in $\mathfrak{D}$ as $-E_{ij}$. We conclude that any representation σ_V of the Lie k-algebra $\mathfrak{gl}(\mathfrak{D})$ in a k-module V gives rise to a representation ρ of the Lie k-algebra $R\mathfrak{D}$ in $R \otimes_k V$ defined by the formula

$$\rho(f \otimes \partial_i) = f\partial_i \otimes \mathrm{id}_V - \sum_{j=1}^{n} \partial_j f \otimes \sigma_V(E_{ij}), \qquad f \in R.$$

In particular, to the natural representation of $\mathfrak{gl}(\mathfrak{D})$ in $\mathfrak{D}$ there corresponds none other than the adjoint representation of $R\mathfrak{D}$. Another special case of a trivial $\mathfrak{gl}(\mathfrak{D})$-module k, when $\sigma_k = 0$, produces the natural representation of $R\mathfrak{D}$ in $\mathcal{R}k \cong R$.

Our main purpose is to study the submodules of the modules $\mathcal{R}V$ and the homomorphisms between two modules of such a kind. In doing this it is natural to work in a certain category of representations which we now introduce.

Definition. Denote by $\mathcal{C}$ the category whose objects are additive groups M endowed with a structure of an R-module, a structure ρ_M of an $R\mathfrak{D}$-module and a structure σ_M of a discrete $\widehat{W}_n(k)_0$-module so that the following properties are satisfied:

(1.1) $[\rho_M(D), f_M] = (Df)_M$,

(1.2) $[\sigma_M(D'), f_M] = 0$,

(1.3) $[\rho_M(\partial_i), \sigma_M(D')] = 0$,

(1.4) $\rho_M(f\partial_i) = f_M \circ \rho_M(\partial_i) + \sum_{\alpha>0} (\partial^\alpha f)_M \circ \sigma_M(x^{(\alpha)}\partial_i)$

for all $f \in R$, $D \in R\mathfrak{D}$, $D' \in \widehat{W}_n(k)_0$, $i = 1, \ldots, n$, where for $g \in R$ we let g_M denote the corresponding operator on M afforded by the R-module structure. The morphisms in $\mathcal{C}$ are the mappings that respect the three module structures.

In general, given a Lie k-algebra L operating on R as a Lie algebra of derivations, we call an additive group M endowed with an R-module structure and an L-module structure ρ_M an (R, L)-module if (1.1) holds for all $D \in L$ and $f \in R$. Thus we may regard any $M \in \mathcal{C}$ as an $(R, R\mathfrak{D})$-module. If M, N are (R, L)-modules then $M \oplus N$, $M \otimes_R N$ and $\mathrm{Hom}_R(M, N)$ have canonical L-module structures compatible with their R-module structures. That the action of L in the tensor product and in the module of homomorphisms is well defined is verified, for instance, in [22, sect. 1]. If $M, N \in \mathcal{C}$ then there are also $\widehat{W}_n(k)_0$-module structures, and it is immediate from the definitions that $M \oplus N$ and $M \otimes_R N$ are objects of $\mathcal{C}$. For $\mathrm{Hom}_R(M, N)$ this is valid only under the assumption that $\sigma_M(\widehat{W}_n(k)_l) = 0$ and $\sigma_N(\widehat{W}_n(k)_l) = 0$ for some l, since otherwise the representation of $\widehat{W}_n(k)_0$ needn't be discrete. If $M \in \mathcal{C}$ and M' is its additive subgroup closed under the three module structures then M' and M/M' are objects of $\mathcal{C}$ in a natural way. In this case we call M' a $\mathcal{C}$-submodule of M. The objects $M \in \mathcal{C}$ with $\sigma_M = 0$ may be identified with the $(R, R\mathfrak{D})$-modules satisfying $\rho_M(fD) = f_M \circ \rho_M(D)$ for all $f \in R$, $D \in R\mathfrak{D}$. Such $(R, R\mathfrak{D})$-modules were called in [22] connected because they correspond to flat connections on vector bundles in geometry.

For a discrete $\widehat{W}_n(k)_0$-module V we have defined an $R\mathfrak{D}$-module structure on $R \otimes_k V$. There is also a natural R-module structure, and extending the endomorphisms $\sigma_V(D')$ by R-linearity to the whole of $R \otimes_k V$, we see immediately that (1.1)–(1.4) are satisfied for $M = R \otimes_k V$. We thus may view $\mathcal{R}$ as a functor from the category of discrete $\widehat{W}_n(k)_0$-modules to the category $\mathcal{C}$. Forgetting the R-module and the $\widehat{W}_n(k)_0$-module structures, we also view $\mathcal{R}$ as a functor to the category of $R\mathfrak{D}$-modules. The results in Section 4 show that some information about the two forgotten structures can be regained from the $R\mathfrak{D}$-module structure. If V, V' are two discrete $\widehat{W}_n(k)_0$-modules then there are obvious canonical isomorphisms in $\mathcal{C}$

$$\mathcal{R}V \oplus \mathcal{R}V' \cong \mathcal{R}(V \oplus V') \qquad \text{and} \qquad \mathcal{R}V \otimes_R \mathcal{R}V' \cong \mathcal{R}(V \otimes_k V').$$

Remark. If $R = \widehat{O}_n(k)$, and so $R\mathfrak{D} = \widehat{W}_n(k)$, one should distinguish the representation σ_M from the restriction of ρ_M to $\widehat{W}_n(k)_0$.

2. A realization of induced and coinduced modules

We are going to relate the constructed modules $\mathcal{R}V$ to the coinduced ones. For this purpose we need an algebraic interpretation of the concept of jets familiar in differential geometry. To employ the geometric intuition we wish to consider points with values in ring extensions of the ground ring k. The theory of k-functors, i. e. functors from the category of commutative (associative and unital) k-algebras to the category of sets, provides the necessary background. The k-algebra R determines an affine scheme $\mathfrak{Sp}\, R$ which is the k-functor that associates to each commutative k-algebra K the set of all k-algebra homomorphisms $R \to K$. By a K-valued point we understand a K-valued point of this functor, i. e. an arbitrary homomorphism $R \to K$.

Let M be an $(R, \mathfrak{D})$-module and $\xi : R \to K$ a homomorphism. Denote by $\xi_* M$ the K-module obtained from M by means of extension of scalars. In other words, $\xi_* M = K \otimes_R M$ where K is regarded as an R-algebra via ξ. Define a mapping $i_\xi : M \to \xi_* M$ by the rule $i_\xi(m) = 1 \otimes m$ for $m \in M$. Let $\mathfrak{D}_K$ designate the commutative Lie K-algebra with the underlying K-module $K \otimes_k \mathfrak{D}$. Its universal enveloping algebra $U(\mathfrak{D}_K)$ is canonically isomorphic with the extension $U(\mathfrak{D})_K$ of the universal enveloping algebra of $\mathfrak{D}$. We define the K-module of jets of infinite order associated with M at the K-valued point ξ as follows:

$$\mathfrak{J}_\xi(M) = \mathrm{Hom}_k(U(\mathfrak{D}), \xi_* M) \cong \mathrm{Hom}_K(U(\mathfrak{D}_K), \xi_* M).$$

To an element $m \in M$ there corresponds its jet $j_\xi(m) \in \mathfrak{J}_\xi(M)$ such that

$j_\xi(m)(u) = i_\xi(um)$ for $u \in U(\mathfrak{D})$. The Lie algebra $\mathfrak{D}$ operates in $\mathfrak{J}_\xi(M)$ by the rule $(D\varphi)(u) = \varphi(uD)$ for $D \in \mathfrak{D}$, $\varphi \in \mathfrak{J}_\xi(M)$, $u \in U(\mathfrak{D})$. Obviously, j_ξ is a $\mathfrak{D}$-module homomorphism.

Consider further the particular case $M = \mathcal{R}V = R \otimes_k V$. The $\mathfrak{D}$-module structure on M is given by the action of $\mathfrak{D}$ on the first factor. Here $\xi_* M \cong V_K = K \otimes_k V$ is a discrete module for the Lie K-algebra $K \otimes_k \widehat{W}_n(k)_0$. By continuity the Lie K-algebra $\widehat{W}_n(K)_0 \cong K \widehat{\otimes}_k \widehat{W}_n(k)_0$ also operates on V_K. Now $\mathfrak{D}_K$ is identified with a subalgebra in $\widehat{W}_n(K)$ such that $\widehat{W}_n(K) = \mathfrak{D}_K \oplus \widehat{W}_n(K)_0$. It follows that the elements ∂^α, $\alpha \in A(n)$, constitute a basis of the universal enveloping algebra $U(\widehat{W}_n(K))$ as a left regular module for its subalgebra $U(\widehat{W}_n(K)_0)$. Although we can not refer to the Poincaré–Birkhoff–Witt theorem to deduce this fact, because $\widehat{W}_n(K)$ is not a free K-module in general, quite standard verifications allow us to construct a right $U(\widehat{W}_n(K))$-module structure on $U(\widehat{W}_n(K)_0) \otimes_K U(\mathfrak{D}_K)$ such that the actions of the elements $D \in \widehat{W}_n(K)$ are given by the rule

$$(u \otimes \partial^\alpha) \cdot D = \sum_{0 \le \beta \le \alpha} \binom{\alpha}{\beta} \left(u \cdot \mathrm{pr}\big((\mathrm{ad}\,\partial)^\beta D\big) \otimes \partial^{\alpha-\beta} + u \otimes \mathrm{pr}'\big((\mathrm{ad}\,\partial)^\beta D\big) \cdot \partial^{\alpha-\beta}\right)$$

for $u \in U(\widehat{W}_n(K)_0)$, $\alpha \in A(n)$, where $\mathrm{pr} : \widehat{W}_n(K) \to \widehat{W}_n(K)_0$ and $\mathrm{pr}' : \widehat{W}_n(K) \to \mathfrak{D}_K$ are the projections associated with the decomposition of $\widehat{W}_n(K)$ written above. Given a family of elements $u_\alpha \in U(\widehat{W}_n(K)_0)$, $\alpha \in A(n)$, only a finite number of which are distinct from zero, we have $(1 \otimes 1) \cdot \sum u_\alpha \partial^\alpha = \sum u_\alpha \otimes \partial^\alpha$. It follows that the elements ∂^α, $\alpha \in A(n)$, are linearly independent over $U(\widehat{W}_n(K)_0)$, whence the claim is clear. We see that each K-linear mapping $U(\mathfrak{D}_K) \to V_K$ extends uniquely to a $U(\widehat{W}_n(K)_0)$-module homomorphism $U(\widehat{W}_n(K)) \to V_K$. Thus $\mathfrak{J}_\xi(\mathcal{R}V)$ has a canonical structure of a $\widehat{W}_n(K)$-module coinduced from the $\widehat{W}_n(K)_0$-module V_K:

$$\mathfrak{J}_\xi(\mathcal{R}V) \cong \mathrm{Coind}\, V_K = \mathrm{Hom}_{U(\widehat{W}_n(K)_0)}(U(\widehat{W}_n(K)),\, V_K).$$

Interpreting an element $\varphi \in \mathfrak{J}_\xi(\mathcal{R}V)$ as a mapping $U(\widehat{W}_n(K)) \to V_K$, we have $(D\varphi)(u) = \varphi(uD)$ for $D \in \widehat{W}_n(K)$, $u \in U(\widehat{W}_n(K))$.

As the algebra R is an arbitrary one, we may apply the preceding deliberations to the K-algebra $\widehat{O}_n(K)$. Let $\pi : \widehat{O}_n(K) \to K$ be the K-algebra homomorphism such that $\pi(x^{(\alpha)}) = 0$ for $\alpha > 0$. There is the corresponding mapping

$$j_\pi : \widehat{O}_n(K) \otimes_K V_K \to \mathrm{Coind}\, V_K .$$

We can compute it explicitly. Let $f = \sum \lambda_\alpha x^{(\alpha)} \in \widehat{O}_n(K)$, where $\lambda_\alpha \in K$, and let $v \in V_K$. Since $\pi(\partial^\beta x^{(\alpha)}) = \delta_{\alpha\beta}$, we have $j_\pi(f \otimes v)(\partial^\beta) = \lambda_\beta v$. It is immediate from this description that j_π is a K-module isomorphism whenever V is free of finite rank as a k-module. In particular, taking the trivial module k, and next the module $\mathfrak{D} \cong \widehat{W}_n(k)/\widehat{W}_n(k)_0$ for V, we get K-module isomorphisms

$$\widehat{O}_n(K) \cong \mathrm{Coind}\, K \qquad \text{and} \qquad \widehat{W}_n(K) \cong \mathrm{Coind}\, \mathfrak{D}_K .$$

Returning to the general case, we get the composite mappings

$$j_\xi : R \to \mathfrak{J}_\xi(R) \cong \mathrm{Coind}\, K \cong \widehat{O}_n(K),$$

$$j_\xi : R\mathfrak{D} \to \mathfrak{J}_\xi(R\mathfrak{D}) \cong \operatorname{Coind}\mathfrak{D}_K \cong \widehat{W}_n(K)$$

for which we retain the notation j_ξ. Let $f \in R$. Then $j_\xi(f \otimes \partial_i)$ can be written as $\sum_j \sum_\alpha (\lambda_{\alpha j} x^{(\alpha)})\partial_j$ with certain $\lambda_{\alpha j} \in K$. For each $\alpha \in A(n)$ we have $\sum_j \lambda_{\alpha j}\partial_j = j_\xi(f\otimes\partial_i)(\partial^\alpha) = i_\xi((\operatorname{ad}\partial)^\alpha(f\otimes\partial_i)) = i_\xi(\partial^\alpha f \otimes \partial_i) = \xi(\partial^\alpha f)\partial_i$. Thus

$$j_\xi(f\otimes\partial_i) = \sum_\alpha \xi(\partial^\alpha f)x^{(\alpha)}\partial_i\,, \qquad f \in R.$$

Now we are in a position to relate the action of $R\mathfrak{D}$ in $\mathcal{R}V$ defined in Section 1 to that of $\widehat{W}_n(K)$ in the coinduced module. We claim that

$$j_\xi(\rho(D)m) = j_\xi(D)\cdot j_\xi(m) \qquad \text{for all} \quad D \in R\mathfrak{D}\,,\ m \in \mathcal{R}V\,.$$

Denote $U' = \{u \in U(\mathfrak{D}) \mid j_\xi(\rho(D)m)(u) = (j_\xi(D)\cdot j_\xi(m))(u)$ for all $D \in R\mathfrak{D}$, $m \in \mathcal{R}V\}$. We have to prove that $U' = U(\mathfrak{D})$. It is immediate from the definition of the endomorphisms $\rho(D)$ that $\rho([\partial_i, D]) = [\rho(\partial_i), \rho(D)]$. Noting that the mappings j_ξ commute with the actions of $\mathfrak{D}$, we get

$$\begin{aligned} j_\xi(\rho(D)m)(u\partial_i) &= \big(\partial_i \cdot j_\xi(\rho(D)m)\big)(u)\\ &= j_\xi(\rho([\partial_i, D])m)(u) + j_\xi(\rho(D)\rho(\partial_i)m)(u)\,;\end{aligned}$$

$$\begin{aligned} (j_\xi(D)\cdot j_\xi(m))(u\partial_i) &= \big(\partial_i\cdot(j_\xi(D)\cdot j_\xi(m))\big)(u)\\ &= \big(j_\xi([\partial_i, D])\cdot j_\xi(m)\big)(u) + \big(j_\xi(D)\cdot j_\xi(\rho(\partial_i)m)\big)(u)\,.\end{aligned}$$

Hence $u\partial_i \in U'$ whenever $u \in U'$, i. e. U' is a right ideal. So it remains to compare the values at $u = 1$. Let $D = f\otimes\partial_i$, $m = g\otimes v$. Then

$$\begin{aligned} \big(j_\xi(D)\cdot j_\xi(m)\big)(1) &= j_\xi(m)\big(j_\xi(D)\big) = j_\xi(m)\big(\textstyle\sum_\alpha \xi(\partial^\alpha f)x^{(\alpha)}\partial_i\big)\\ &= \xi(f)j_\xi(m)(\partial_i) + \sigma_V\big(\textstyle\sum_{\alpha>0}\xi(\partial^\alpha f)x^{(\alpha)}\partial_i\big)j_\xi(m)(1)\\ &= \xi(f)\xi(\partial_i g)v + \sigma_V\big(\textstyle\sum_{\alpha>0}\xi(\partial^\alpha f)x^{(\alpha)}\partial_i\big)(\xi(g)v)\\ &= i_\xi(\rho(D)m) = j_\xi(\rho(D)m)(1)\,,\end{aligned}$$

as required.

As a particular case take $V = \mathfrak{D}$. Then the action of $R\mathfrak{D}$ on $\mathcal{R}V = R\mathfrak{D}$ is the adjoint one. Hence $j_\xi([D, D']) = j_\xi(D)\cdot j_\xi(D')$. This is valid also for the K-algebra $\widehat{O}_n(K)$ instead of R, so that the adjoint representation of $\widehat{W}_n(K)$ corresponds to its representation in the coinduced module $\operatorname{Coind}\mathfrak{D}_K$ under isomorphism j_π. It follows

$$j_\xi([D, D']) = [j_\xi(D), j_\xi(D')] \qquad \text{for all} \quad D, D' \in R\mathfrak{D}\,.$$

Let V, V' be two discrete $\widehat{W}_n(k)_0$-modules. As mentioned in Section 1, there is a multiplication $\mathcal{R}V \otimes_k \mathcal{R}V' \to \mathcal{R}(V\otimes_k V')$ such that $(f\otimes v)\otimes(g\otimes v')$ is sent to $fg\otimes(v\otimes v')$ for $f, g \in R$, $v \in V$, $v' \in V'$. There is a similar multiplication $\operatorname{Coind} V_K \otimes_K \operatorname{Coind} V'_K \to \operatorname{Coind}(V_K \otimes_K V'_K)$ introduced in [1]. The latter is characterized by the properties that it is $\widehat{W}_n(K)$-invariant and satisfies

$(\varphi\psi)(1) = \varphi(1)\psi(1)$ for $\varphi \in \operatorname{Coind} V$, $\psi \in \operatorname{Coind} V'$. The two multiplications are compatible, i. e.

$$j_\xi(m\,m') = j_\xi(m)\,j_\xi(m') \qquad \text{for} \quad m \in \mathfrak{R}V\,,\ m' \in \mathfrak{R}V'\,.$$

In view of the $\mathfrak{D}$-invariance of the multiplications it suffices to verify only that $j_\xi(m\,m')(1) = (j_\xi(m)\,j_\xi(m'))(1)$. Putting $m = f \otimes v$, $m' = g \otimes v'$, one transforms both sides to $\xi(f)\xi(g)\,v \otimes v'$, as required.

The multiplications on coinduced modules provide $\operatorname{Coind} K$ with a structure of a commutative algebra, $\operatorname{Coind} V_K$ with that of a module over $\operatorname{Coind} K$. We see that $j_\xi : R \to \operatorname{Coind} K$ is an algebra homomorphism and the R-module structure on $\mathfrak{R}V$ is compatible with the $\operatorname{Coind} K$-module structure on $\operatorname{Coind} V_K$. This assertion includes the well-known fact that $j_\pi : \widehat{O}_n(K) \to \operatorname{Coind} K$ is an isomorphism of K-algebras. So $\operatorname{Coind} V_K$ has a structure of an $\widehat{O}_n(K)$-module compatible with that of a $\widehat{W}_n(K)$-module.

Now take $K = R$ and $\xi = \mathrm{id}_R$. For an arbitrary module M and an element $m \in M$ we have $\xi_* M \cong M$ and $j_\xi(m)(1) = i_\xi(m) = m$. So the mappings j_ξ are injective in this case.

Let us summarize the main results of this section.

Theorem 2.1. *There is a canonical embedding of $R\mathfrak{D}$ as a Lie k-subalgebra into the Lie R-algebra $\widehat{W}_n(R)$. There is a canonical embedding of $\mathfrak{R}V$ into the $\widehat{W}_n(R)$-module* $\operatorname{Coind} V_R$ *coinduced from the $\widehat{W}_n(R)_0$-module V_R. The action of $R\mathfrak{D}$ on $\mathfrak{R}V$ defined in Section* 1 *is the restriction of its action on the coinduced module.*

So far we have not used the fact that $\rho([D, D']) = [\rho(D), \rho(D')]$ for all $D, D' \in R\mathfrak{D}$. Since the embedding $R\mathfrak{D} \to \widehat{W}_n(R)$ gives a representation of $R\mathfrak{D}$ in $\operatorname{Coind} V_R$, it is obvious now that the endomorphisms $\rho(D)$ do define a representation of $R\mathfrak{D}$ in $\mathfrak{R}V$.

Theorem 2.2. *Let $R = \widehat{O}_n(k)$. Assume that V is free of finite rank as a k-module. Then there is a canonical isomorphism of $(\widehat{O}_n(k), \widehat{W}_n(k))$-modules $\mathfrak{R}V \cong$* $\operatorname{Coind} V$.

Remarks. For a k-submodule $E \subseteq U(\mathfrak{D})$ denote $E^\perp = \{\varphi \in \operatorname{Coind} V \mid \varphi(E) = 0\}$. Endow $\operatorname{Coind} V$ with the linear topology whose neighbourhoods of zero are those subsets that contain some $E^\perp$ with a finitely generated E. If V is free of infinite rank over k then $\mathfrak{R}V$ is isomorphic with a dense $\widehat{W}_n(k)$-submodule of $\operatorname{Coind} V$.

Blattner's paper [1] contains a characterization of coinduced modules. In particular, if M is an $(\widehat{O}_n(k), \widehat{W}_n(k))$-module which is free of finite rank over R then M is complete and separated with respect to the topology determined by the filtration $\widehat{\mathfrak{m}}^{(l)}M$, $l \geq 0$, whence there is a canonical isomorphism $M \cong \operatorname{Coind}(M/\widehat{\mathfrak{m}}M)$. Although [1] deals with the Lie algebras over a field only, this result is valid for an arbitrary commutative ring k. So Theorem 2.2 is a consequence of Blattner's theorem as soon as we know that ρ is actually a representation of $\widehat{W}_n(k)$. What is more important, Theorem 2.2 gives an explicit formula for the action of $\widehat{W}_n(k)$ in the coinduced modules.

The induced $\widehat{W}_n(k)$-modules $\operatorname{Ind} V = U(\widehat{W}_n(k)) \otimes_{U(\widehat{W}_n(k)_0)} V$ admit a similar decription which involves, however, a certain twisting in the category $\mathcal{C}$. This description is related to the multiplications of a different kind. Given two $\widehat{W}_n(k)_0$-modules V, V' there is an obvious canonical $\widehat{W}_n(k)_0$-module homomorphism $V \to \operatorname{Hom}_k(V', V \otimes_k V')$. The canonical $\widehat{W}_n(k)_0$-module homomorphisms $\operatorname{Coind} V' \to V'$ and $V \otimes_k V' \to \operatorname{Ind}(V \otimes_k V')$ determine by functoriality a $\widehat{W}_n(k)_0$-module homomorphism $\operatorname{Hom}_k(V', V \otimes_k V') \to \operatorname{Hom}_k(\operatorname{Coind} V', \operatorname{Ind}(V \otimes_k V'))$, hence by the universality property of the induced modules a certain $\widehat{W}_n(k)$-module homomorphism $\nu : \operatorname{Ind} V \to \operatorname{Hom}_k(\operatorname{Coind} V', \operatorname{Ind}(V \otimes_k V'))$. We put $q \leftharpoondown \varphi = \nu(q)(\varphi)$ for $q \in \operatorname{Ind} V$, $\varphi \in \operatorname{Coind} V'$. Explicitly, denoting by Δ the comultiplication, by S the antipode of the Hopf algebra $U(\widehat{W}_n(k))$, we have $(u \otimes v) \leftharpoondown \varphi = \sum_i u_i' \otimes \big(v \otimes \varphi(Su_i'')\big)$ for $u \in U(\widehat{W}_n(k))$, $v \in V$, $\varphi \in \operatorname{Coind} V'$, where we put $\Delta u = \sum_i u_i' \otimes u_i''$. The assignment $q \otimes \varphi \mapsto q \leftharpoondown \varphi$ defines a $\widehat{W}_n(k)$-invariant multiplication

$$\operatorname{Ind} V \otimes_k \operatorname{Coind} V' \to \operatorname{Ind}(V \otimes_k V').$$

Given three $\widehat{W}_n(k)_0$-modules V, V', V'', we have $q \leftharpoondown (\varphi\psi) = (q \leftharpoondown \varphi) \leftharpoondown \psi$ for $q \in \operatorname{Ind} V$, $\varphi \in \operatorname{Coind} V'$, $\psi \in \operatorname{Coind} V''$. In particular, $\operatorname{Ind} V$ carries a canonical structure of a module over $\operatorname{Coind} k \cong \widehat{O}_n(k)$ and the multiplication above is $\widehat{O}_n(k)$-balanced.

Let further $R = \widehat{O}_n(k)$ and let V' be a discrete $\widehat{W}_n(k)_0$-module. We put $q \leftharpoondown m = q \leftharpoondown j_\pi(m)$ for $q \in \operatorname{Ind} V$, $m \in \mathcal{R}V'$. Then $(u \otimes v) \leftharpoondown (1 \otimes v') = u \otimes (v \otimes v')$ for $u \in U(\mathfrak{D})$, $v \in V$, $v' \in V'$, and we conclude that

$$\operatorname{Ind} V \otimes_R \mathcal{R}V' \cong \operatorname{Ind}(V \otimes_k V').$$

The n-th exterior powers $\bigwedge^n \mathfrak{D}$ and $\bigwedge^n \mathfrak{D}^*$ of the k-module $\mathfrak{D}$ and of its dual $\mathfrak{D}^*$ are free k-modules of rank 1 and $\bigwedge^n \mathfrak{D} \otimes_k \bigwedge^n \mathfrak{D}^* \cong k$. The $\widehat{W}_n(k)_0$-module structure on $\mathfrak{D}$ determines those on the exterior powers. By the above

$$\operatorname{Ind} V \cong \operatorname{Ind}(\textstyle\bigwedge^n \mathfrak{D}) \otimes_R \mathcal{R}(\bigwedge^n \mathfrak{D}^* \otimes_k V).$$

Put $Q = \operatorname{Ind}(\bigwedge^n \mathfrak{D})$. The dual $\widehat{W}_n(k)$-module $Q^* = \operatorname{Hom}_k(Q, k)$ has also an R-module structure defined by the rule $\langle f\omega, q\rangle = \langle \omega, fq\rangle$ for $f \in R$, $\omega \in Q^*$, $q \in Q$. By the duality between the induced and the coinduced modules there is a $\widehat{W}_n(k)$-module isomorphism $Q^* \cong \operatorname{Coind}(\bigwedge^n \mathfrak{D}^*)$, and one sees that it respects the R-module structures as well. Now we may identify Q^* with the module $\widehat{\Omega}^n$ of differential forms of degree n. As is well known, $(fD)\omega = D(f\omega)$ for all $f \in R$, $D \in \widehat{W}_n(k)$, $\omega \in \widehat{\Omega}^n$. We deduce from the formula

$$\langle \omega, (fD)q\rangle = \langle -(fD)\omega, q\rangle = \langle -D(f\omega), q\rangle = \langle \omega, f(Dq)\rangle$$

that $(fD)q = f(Dq)$ for all $f \in R$, $D \in \widehat{W}_n(k)$, $q \in Q$. If $\eta : Q \to Q$ is an $(R, \mathfrak{D})$-module endomorphism, then its transpose $\eta^* : Q^* \to Q^*$ is also an $(R, \mathfrak{D})$-module endomorphism. Since $\widehat{\Omega}^n$ is free of rank 1 as a module over R, the endomorphism η^* is the multiplication by an element $\lambda \in R$ which is annihilated by $\mathfrak{D}$, i. e. $\lambda \in k$. It follows that η is the multiplication by λ as well. We use here that Q is free over k, since its underlying k-module is isomorphic with $U(\mathfrak{D})$. If k is a field, then R is a $\mathfrak{D}$-simple algebra. Hence $\widehat{\Omega}^n$ contains no nonzero proper $\mathfrak{D}$-invariant R-submodules. Now if Q_1 is a $\mathfrak{D}$-invariant R-submodule of Q, then $Q_1^\perp = \{\omega \in \widehat{\Omega}^n \mid \langle\omega, Q_1\rangle = 0\}$ is a $\mathfrak{D}$-invariant R-submodule of Q^*, and it follows that $Q_1 = 0$ or $Q_1 = Q$. We can state our result:

Theorem 2.3. *Let $R = \widehat{O}_n(k)$. Then $\operatorname{Ind} V \cong Q \otimes_R \mathcal{R}(\bigwedge^n \mathfrak{D}^* \otimes_k V)$ where Q is an object of $\mathcal{C}$ with $\sigma_Q = 0$ and $\operatorname{End}_{\mathcal{C}} Q \cong k$. If k is a field then Q is a simple object of $\mathcal{C}$.*

3. Independence of differential operators

We shall study here the concept of independence of operators. Propositions 3.4, 3.5 give us a tool for the investigation of the structure of modules in Section 4.

Endow the endomorphism algebra $\operatorname{End}_k R$ with an R-module structure by putting $(f\varphi)(g) = f\,\varphi(g)$ for $f, g \in R$, $\varphi \in \operatorname{End}_k R$. There is a notion of linear independence of endomorphisms over R with respect to this structure. We shall need a stronger property.

Definition. We say that a system of endomorphisms $\Phi \subseteq \operatorname{End}_k R$ is independent if for any finite subset $\Phi' = \{\varphi_1, \ldots, \varphi_n\} \subseteq \Phi$ the submodule $\operatorname{Val}\Phi'$ of the free R-module R^n generated by all n-tuples $(\varphi_1(g), \ldots, \varphi_n(g))$ with $g \in R$ coincides with R^n.

Obviously, any subsystem of an independent system of endomorphisms is itself independent. Now, the linear dependence $f_1\varphi_1 + \cdots + f_n\varphi_n = 0$ with coefficients $f_1, \ldots, f_n \in R$ means just that the R-module homomorphism $\eta : R^n \to R$ defined by the rule $\eta(g_1, \ldots, g_n) = \sum f_i g_i$ for $g_1, \ldots, g_n \in R$ vanishes on $\operatorname{Val}\Phi'$. The equality $\operatorname{Val}\Phi' = R^n$ implies $\eta = 0$, i. e. $f_1 = \cdots = f_n = 0$. Thus the independence implies the linear independence over R. The converse is not true in general. However, we can give some sufficient conditions. Recall that R is said to be differentially simple with respect to a system of its derivations N if R contains no nonzero proper N-invariant ideals.

Proposition 3.1. *Let $\Phi \subseteq \operatorname{End}_k R$ and $N \subseteq \operatorname{Der} R$. Assume that R is N-simple. Then Φ is an independent system in both of the following two cases:*

(i) *the endomorphisms in Φ are linearly independent over R and each $\varphi \in \Phi$ is contained in a finite subset $\Phi' \subseteq \Phi$ such that N normalizes the R-module $R\Phi'$,*

(ii) *the endomorphisms in Φ are linearly independent over the subring $R^N = \{f \in R \mid Df = 0$ for all $D \in N\}$ (which is actually a field) of R and $[N, \Phi] = 0$.*

Proof. (i) By the hypotheses any finite subset of Φ is contained in another finite subset Φ' with the property that $[N, R\Phi'] \subseteq R\Phi'$. So it will suffice to prove the equality $\operatorname{Val}\Phi' = R^n$, where n is the cardinality of Φ', for such a Φ'. The R-module $M = R\Phi'$ is free with Φ' as its basis. Identify $\operatorname{Val}\Phi'$ with the submodule of the dual R-module $M^* = \operatorname{Hom}_R(M, R)$ generated by the homomorphisms ξ_g, $g \in R$, such that $\xi_g(\varphi) = \varphi(g)$ for $\varphi \in M$. Denote by $\widetilde{N}$ the normalizer of M in $\operatorname{Der} R$. Then $\widetilde{N}$ is a Lie subring of $\operatorname{Der} R$ containing N. In particular, R is $\widetilde{N}$-simple. We may regard M, hence also M^* as $(R, \widetilde{N})$-modules. An immediate verification shows that $D\xi_g = \xi_{Dg}$ for $D \in \widetilde{N}$, $g \in R$. Hence $\operatorname{Val}\Phi'$ is stable under the action of $\widetilde{N}$. We see that the factor $P = M^*/\operatorname{Val}\Phi'$ carries the induced structure of an $(R, \widetilde{N})$-module. Now [22, Th. 1.6] ensures that P is projective as an R-module. The property of linear independence of Φ' implies that there are no nontrivial R-module homomorphisms $M^* \to R$ vanishing on $\operatorname{Val}\Phi'$, i. e. $\operatorname{Hom}_R(P, R) = 0$. It follows $P = 0$, i. e. $\operatorname{Val}\Phi' = M^*$, as required.

(ii) We use the arguments similar to those in [11] to show that the endomorphisms in Φ are linearly independent over R and then apply (i). Assume that $\varphi_1, \ldots, \varphi_n$ are linearly dependent over R and that all subsystems of Φ of less cardinality are linearly independent over R. Then $I = \{f \in R \mid f\varphi_n \in R\varphi_1 + \cdots + R\varphi_{n-1}\}$ is a nonzero ideal of R. If $f \in I$ then $(Df)\varphi_n = [D, f\varphi_n] \in R\varphi_1 + \cdots + R\varphi_{n-1}$, i. e. $Df \in I$ for any $D \in N$. Hence I is stable under N, and so $I = R$ in view of N-simplicity of R. We have $\varphi_n = f_1\varphi_1 + \cdots + f_{n-1}\varphi_{n-1}$ for suitable $f_1, \ldots, f_{n-1} \in R$. If $D \in N$ then $(Df_1)\varphi_1 + \cdots + (Df_{n-1})\varphi_{n-1} = [D, \varphi_n] = 0$, whence $Df_1 = \cdots = Df_{n-1} = 0$ since $\varphi_1, \ldots, \varphi_{n-1}$ are linearly independent over R. It follows $f_1, \ldots, f_{n-1} \in R^N$, which contradicts the linear independence of $\varphi_1, \ldots, \varphi_n$ over R^N.

Next we study the properties of an independent system of derivations which we needn't assume to be pairwise commuting for the moment.

Proposition 3.2. *Let $\partial_1, \ldots, \partial_n$ be independent derivations of R. Assume that s is a positive integer such that $s!\,R = R$. Then the system of differential operators $\partial^\alpha = \partial_1^{\alpha_1} \ldots \partial_n^{\alpha_n}$, $\alpha \in A'$, where $A' = \{\alpha = (\alpha_1, \ldots, \alpha_n) \in A(n) \mid \alpha_i \leq s$ for all $i = 1, \ldots, n\}$, is independent.*

Proof. Let $\alpha^1, \ldots, \alpha^l$ be distinct elements of A'. Proceeding by induction on $a = \max_{1\leq i\leq l} |\alpha^i|$, we shall prove that $\operatorname{Val}(\partial^{\alpha^1}, \ldots, \partial^{\alpha^l}) = R^l$. If $a = 0$ then $l = 1$ and $\alpha_1 = 0$, so that the assertion is clear. Let $a > 0$. Denote $M = \operatorname{Val}(\partial^{\alpha^1}, \ldots, \partial^{\alpha^l})$, and let $e_1, \ldots, e_l$ be the standard basis of the R-module R^l. We have $\sum_t \partial^{\alpha^t}(g)e_t \in M$ for all $g \in R$. If $\alpha^u = 0$ for some u, then putting $g = 1$, we get $e_u \in M$ at once. So the assertion is reduced to the case that $0 \notin \{\alpha^1, \ldots, \alpha^l\}$. Let us assume this. For $f, g \in R$ we have

$$\begin{aligned}\zeta(f, g) &= \sum_t \partial^{\alpha^t}(fg)\, e_t - f \sum_t \partial^{\alpha^t}(g)\, e_t - g \sum_t \partial^{\alpha^t}(f)\, e_t \\ &= \sum_t \Big(\sum_{0<\beta<\alpha^t} \binom{\alpha^t}{\beta} \partial^\beta(f)\, \partial^{\alpha^t-\beta}(g) \Big)\, e_t \in M.\end{aligned}$$

Next we shall apply the induction hypotheses to the finite set $A'' = \{\alpha \in A(n) \mid 0 < \alpha < \alpha^t$ for some $t = 1, \ldots, l\} \subseteq A'$. Fixing $\gamma, \delta \in A''$, we can choose therefore a finite number of elements f'_μ, f''_μ, g'_ν, $g''_\nu \in R$ such that

$$\sum_\mu f'_\mu \partial^\beta f''_\mu = \begin{cases} 1 \text{ for } \beta = \gamma \\ 0 \text{ for } \beta \in A'',\ \beta \neq \gamma, \end{cases} \qquad \sum_\nu g'_\nu \partial^\beta g''_\nu = \begin{cases} 1 \text{ for } \beta = \delta \\ 0 \text{ for } \beta \in A'',\ \beta \neq \delta. \end{cases}$$

Then

$$\sum_{\mu,\nu} f'_\mu g'_\nu \zeta(f''_\mu, g''_\nu) = \sum_{\alpha^t = \gamma+\delta} \binom{\alpha^t}{\gamma} e_t \in M.$$

Since $\alpha^1, \dots, \alpha^l$ are distinct, the sum on the right contains at most one term. For a fixed $u \in \{1, \dots, n\}$ we can write $\alpha^u = \gamma + \delta$ with suitable γ, $\delta \in A''$ provided $|\alpha^u| \geq 2$. Since the coefficient $\binom{\alpha^u}{\gamma}$ is invertible in R, it follows $e_u \in M$ provided $|\alpha^u| \geq 2$. Let $T = \{t \mid |\alpha^t| = 1\}$. We have now $\sum_{t\in T} \partial^{\alpha^t}(f) e_t \in M$ for all $f \in R$. Since the derivations $\partial_1, \dots, \partial_n$ are independent, we deduce $e_u \in M$ for $u \in T$ as well. Thus $M = R^l$.

It may be of interest to note that an analogous assertion is valid for the linear independence:

Proposition 3.3. *Let derivations $\partial_1, \dots, \partial_n$ be linearly independent over R. Let s and A' be the same as in Proposition 3.2. Then the differential operators $\partial^\alpha = \partial_1^{\alpha_1} \dots \partial_n^{\alpha_n}$, $\alpha \in A'$, are linearly independent over R.*

Proof. Suppose that $\sum_{\alpha \in A'} a_\alpha \partial^\alpha = 0$ with coefficients $a_\alpha \in R$ not all of which are zero. Suppose in addition that we have an expression with $\max_{a_\alpha \neq 0} |\alpha|$ minimal possible. We have $a_0 = \sum a_\alpha \partial^\alpha(1) = 0$,

$$\sum a_\alpha \partial^\alpha(fg) - f \sum a_\alpha \partial^\alpha(g) - g \sum a_\alpha \partial^\alpha(f) = \sum_{0<\beta<\alpha} \binom{\alpha}{\beta} a_\alpha \partial^\beta(f) \partial^{\alpha-\beta}(g) = 0.$$

The minimality assumption implies $\sum_{\alpha>\beta} \binom{\alpha}{\beta} a_\alpha \partial^{\alpha-\beta}(g) = 0$ for all $\beta > 0$, and further $\binom{\alpha}{\beta} a_\alpha = 0$ for all α, β such that $0 < \beta < \alpha$. It follows $a_\alpha = 0$ provided $|\alpha| \geq 2$. The linear independence of $\partial_1, \dots, \partial_n$ implies that $a_\alpha = 0$ for $|\alpha| = 1$ as well.

Assume further that k is a field. Let $\operatorname{char} k = p$. If $p > 0$ and $\mathbf{m} = (m_1, \dots, m_n)$, where each m_i is either a positive integer or $+\infty$, denote

$$A(n : \mathbf{m}) = \{\alpha = (\alpha_1, \dots, \alpha_n) \in A(n) \mid \alpha_i < p^{m_i} \text{ for all } i = 1, \dots, n\}.$$

Proposition 3.4. *Suppose that $p = 0$ and $\partial_1, \dots, \partial_n$ is an independent system of derivations of R. Given a finite subset $A' \subseteq A(n)$ and an n-tuple $\gamma \in A'$, there exist a finite number of elements $f_1, \dots, f_u, g_1, \dots, g_u \in R$ such that*

$$(3.1) \quad \sum_{\nu=1}^{u} f_\nu \partial^\alpha g_\nu = \begin{cases} 1 & \text{if } \alpha = \gamma \\ 0 & \text{if } \alpha \in A',\ \alpha \neq \gamma. \end{cases}$$

Proposition 3.5. *Suppose that $p > 0$ and $\{\partial_i^{p^r} \mid 1 \leq i \leq n,\ 0 \leq r < m_i\}$ is an independent system of derivations of R. Given a finite subset $A' \subseteq A(n : \mathbf{m})$ and* *an n-tuple* $\gamma \in A'$, *there exist a finite number of elements* $f_1, \dots, f_u, g_1, \dots, g_u \in R$ *satisfying* (3.1).

Proof. If $p = 0$ then we apply Proposition 3.2 with an arbitrary large s. If $p > 0$ then we apply Proposition 3.2 with $s = p - 1$. It follows that ∂^α, $\alpha \in A(n)$ (respectively $\alpha \in A(n : \mathbf{m})$), are independent endomorphisms of R. It remains just to use the definition of independence.

Remark. Independent systems of derivations are abundant. If, say, $\partial_i x_j = \delta_{ij}$, $1 \le i, j \le n$, for some elements $x_1, \dots, x_n \in R$, then obviously $\partial_1, \dots, \partial_n$ is an independent system. In particular, the partial derivatives of a polynomial algebra, divided power algebra etc. form an independent system of derivations. Proposition 3.2 gives another general example: a system Δ of pairwise commuting derivations is independent if the ring R is differentially simple with respect to Δ and the derivations are linearly independent over the field R^Δ of Δ-invariants. Note also that the preceding results are valid for infinite systems of derivations without any changes in the proofs.

Proposition 3.2 admits an invariant interpretation and also a more conceptual proof. Let C be the coalgebra with a basis ∂^α, $\alpha \in A'$, and the comultiplication $\Delta\, \partial^\alpha = \sum_{0 \le \beta \le \alpha} \binom{\alpha}{\beta} \partial^\beta \otimes \partial^{\alpha-\beta}$. Then $\mathrm{Hom}_k(C, R)$ is an associative k-algebra with respect to convolution operation. Consider the algebra homomorphism $\theta : R \otimes_k R \to \mathrm{Hom}_k(C, R)$ defined by the rule $\theta(f \otimes g)(\partial^\alpha) = f \cdot \partial^\alpha g$ for $f, g \in R$, $\alpha \in A'$. Proposition 3.2 can be restated by saying that the image of θ is dense in $\mathrm{Hom}_k(C, R)$.

4. Submodules and homomorphisms

Let us fix the general assumptions for this section. We assume here that R is a commutative algebra over a field k of characteristic p and $\{\partial_1, \dots, \partial_n\}$ is an independent system of pairwise commuting derivations of R. In case $p > 0$ we fix an n-tuple $\mathbf{m} = (m_1, \dots, m_n)$ with each m_i either a positive integer or $+\infty$ and assume that

the system $\{\partial_i^{p^r} \mid 1 \le i \le n,\ 0 \le r < m_i\}$ of derivations of R is independent;

for each j such that $m_j < \infty$ the derivation $\partial_j^{p^{m_j}}$ is contained in the k-linear span of the subsystem $\{\partial_i^{p^r} \mid 1 \le i \le n,\ 0 \le r < \min(m_i, m_j)\}$.

Denote

$$A'(n : \mathbf{m}) = \{\alpha = (\alpha_1, \dots, \alpha_n) \in A(n) \mid \alpha_i < p^{m_i} - p^{m_i - 1} \text{ for all } i = 1, \dots, n\}.$$

In case $p > 0$, $\mathbf{m} = (\infty, \dots, \infty)$ the further results are completely the same as in case of zero characteristic.

Our results concern objects of the category $\mathcal{C}$ introduced in Section 1. Given objects $M, N \in \mathcal{C}$ and a mapping $\varphi : M \to N$, denote by $\Gamma(\varphi)$ the graph

$\{(m, \varphi m) \mid m \in M\} \subseteq M \oplus N$ of φ. Note that φ respects any of the three module structures if and only if $\Gamma(\varphi)$ is a submodule with respect to the corresponding structure.

Definition. An irreducible height one $\widehat{W}_n(k)_0$-module is called exceptional (respectively coexceptional) if it is isomorphic to $\bigwedge^t(\widehat{W}_n(k)/\widehat{W}_n(k)_0)$ (respectively to $\bigwedge^t(\widehat{W}_n(k)/\widehat{W}_n(k)_0)^*$) for some $t = 0, \ldots, n$.

Proposition 4.1. (i) *Let $M \in \mathcal{C}$. If $p > 0$, assume that*

(4.1) $\sigma_M(x^{(\alpha)}\partial_i) = 0$ *for* $\alpha \in A(n)$ *and* $i = 1, \ldots, n$ *whenever* $\alpha \notin A(n : \mathbf{m})$.

Then any $(R, R\mathfrak{D})$-submodule M' of M is a $\mathcal{C}$-submodule.

(ii) *Let M, $N \in \mathcal{C}$. If $p > 0$, assume that both M and N satisfy* (4.1). *Then any $(R, R\mathfrak{D})$-module homomorphism $\varphi : M \to N$ is a morphism in $\mathcal{C}$.*

Proof. (i) We have to prove that M' is a $\widehat{W}_n(k)_0$-submodule. Given $q \in M'$, there exists $l \geq 0$ such that $\sigma_M(\widehat{W}_n(k)_l)q = 0$. The subset $Q = \{m \in M \mid \sigma_M(\widehat{W}_n(k)_l)m = 0\}$ is a $\widehat{W}_n(k)_0$-submodule of M since $\widehat{W}_n(k)_l$ is an ideal of $\widehat{W}_n(k)_0$. It is an R-submodule and a $\mathfrak{D}$-submodule in view of (1.2), (1.3). The formula (1.4) shows that it is then an $R\mathfrak{D}$-submodule as well, i. e. a $\mathcal{C}$-submodule. Now $Q \cap M'$ is an $(R, R\mathfrak{D})$-submodule of Q. If we show that $Q \cap M'$ is a $\mathcal{C}$-submodule then, since $q \in Q \cap M'$, it will follow $\sigma_M(D')q \in M'$ for all $D' \in \widehat{W}_n(k)_0$.

Thus we have reduced the proof to the case when M is annihilated by $\widehat{W}_n(k)_l$ for some l. Under this assumption the subset $A_i = \{\alpha \in A(n) \mid \alpha > 0$ and $\sigma_M(x^{(\alpha)}\partial_i) \neq 0\}$ is finite for every i. In case of positive characteristic note that $A_i \subseteq A(n : \mathbf{m})$. Applying Propositions 3.4, 3.5 to the subset $A' = A_i \cup \{0\}$ and an n-tuple $\gamma \in A_i$, we find a finite number of elements f_ν, $g_\nu \in R$ satisfying (3.1). We see that M' is stable under the endomorphism

$$\sum_\nu f_\nu \rho_M(g_\nu\partial_i) = \sum_\nu f_\nu \Big(g_\nu \rho_M(\partial_i) + \sum_{\alpha\in A_i} \partial^\alpha(g_\nu)\sigma_M(x^{(\alpha)}\partial_i)\Big) = \sigma_M(x^{(\gamma)}\partial_i).$$

Hence M' is stable under all endomorphisms $\sigma_M(x^{(\alpha)}\partial_i)$, $\alpha > 0$, $i = 1, \ldots, n$.

(ii) The direct sum $M \oplus N$ is an object of $\mathcal{C}$ satisfying (4.1) in case of a positive characteristic. The graph $\Gamma(\varphi)$ is its $(R, R\mathfrak{D})$-submodule. As we have proved already, $\Gamma(\varphi)$ is a $\widehat{W}_n(k)_0$-submodule, whence φ respects the $\widehat{W}_n(k)_0$-module structures.

Theorem 4.2. (i) *Let $M \in \mathcal{C}$. Assume that*

(4.2) *M is completely reducible as a $\widehat{W}_n(k)_0$-module and none of its irreducible height one $\widehat{W}_n(k)_0$-submodules is coexceptional.*

If $p > 0$, assume in addition that

(4.3) $\sigma_M(x^{(\alpha)}\partial_i) = 0$ *for* $\alpha \in A(n)$ *and* $i = 1, \ldots, n$ *whenever* $\alpha \notin A'(n : \mathbf{m})$.

Then any $R\mathfrak{D}$-submodule M' of M is a $\mathcal{C}$-submodule.

(ii) *Let M, N be two objects of $\mathcal{C}$ satisfying* (4.2). *If $p > 0$ assume that both M, N satisfy* (4.3). *Then any $R\mathfrak{D}$-module homomorphism $\varphi : M \to N$ is a morphism in $\mathcal{C}$.*

Proof. Repeating the trick with the graph of φ, we see that (ii) is a consequence of (i). The same arguments as in the proof of Proposition 4.1 reduce the proof of (i) to the case when M is annihilated by $\widehat{W}_n(k)_l$ for some l. In view of Proposition 4.1 it will suffice to verify that M' is an R-submodule of M.

Let $P = \{m \in M \mid Rm \subseteq M'\}$ be the largest R-submodule contained in M' and $Q = RM'$ the smallest R-submodule containing M'. An immediate verification based on (1.1) shows that P, Q are $R\mathfrak{D}$-submodules, hence by Proposition 4.1 $\mathcal{C}$-submodules. We can now pass to an object $Q/P \in \mathcal{C}$ and its $R\mathfrak{D}$-submodule M'/P.

These observations show that we may impose from the very beginning the additional assumptions that M' contains no nonzero R-submodules of M and $RM' = M$. We have to prove that $M = 0$ then. Our general strategy is as follows. We are seeking endomorphisms φ of M lying in the associative subalgebra generated by the endomorphisms $\sigma_M(D')$, $D' \in \widehat{W}_n(k)_0$, with the property that for any $f \in R$ the endomorphism $f\varphi$ belongs to the associative subalgebra generated by the endomorphisms $\rho_M(D)$, $D \in R\mathfrak{D}$. The $R\mathfrak{D}$-submodule M' is stable under $f\varphi$ for any $f \in R$, hence it contains the R-submodule $R\varphi(M')$. By the hypotheses $\varphi(M') = 0$. In view of (1.2) φ is an R-module endomorphism, and so it follows $\varphi(M) = 0$, i. e. $\varphi = 0$. This gives a certain relation between the endomorphisms $\sigma_M(D')$. It appears that there are sufficiently many relations of a similar kind, so that they can not hold simultaneously unless $M = 0$. For brevity we omit further the subscript in the notations ρ_M and σ_M. Let l be the least positive integer such that M is annihilated by $\widehat{W}_n(k)_l$. Consider first the case $l = 1$. Identify σ with a representation of the factor algebra $\widehat{W}_n(k)_0/\widehat{W}_n(k)_1 \cong \mathfrak{gl}(\mathfrak{D})$ as in Section 1. Given a pair of indices j, $j' \in \{1, \ldots, n\}$, let f_ν, $g_\nu \in R$ be a finite number of elements satisfying (3.1) for $A' = \{\alpha \in A(n) \mid |\alpha| \leq 2\}$ and $\gamma = \epsilon_j + \epsilon_{j'}$. For $f \in R$ we have

$$\sum_\nu \partial_s(ff_\nu)g_\nu = \partial_s\Big(\sum_\nu ff_\nu g_\nu\Big) - \sum_\nu ff_\nu\partial_s(g_\nu) = 0,$$

$$\begin{aligned}\sum_\nu \partial_s(ff_\nu)\partial_{s'}(g_\nu) &= \partial_s\Big(\sum_\nu ff_\nu\partial_{s'}(g_\nu)\Big) - \sum_\nu ff_\nu\partial_s\partial_{s'}(g_\nu) \\ &= \begin{cases} 0 & \text{if } \{s, s'\} \neq \{j, j'\} \\ -f & \text{if } \{s, s'\} = \{j, j'\}. \end{cases}\end{aligned}$$

Now

$$\rho(f\partial_i)\,\rho(g\partial_{i'}) = fg\,\rho(\partial_i)\,\rho(\partial_{i'}) + f\partial_i(g)\,\rho(\partial_{i'}) - \sum_{s=1}^{n} \partial_s(f)g\,\sigma(E_{is})\,\rho(\partial_{i'})$$
$$- \sum_{s'=1}^{n} f\partial_{s'}(g)\,\sigma(E_{i's'})\,\rho(\partial_i) - \sum_{s'=1}^{n} f\partial_i\partial_{s'}(g)\,\sigma(E_{i's'})$$
$$+ \sum_{s=1}^{n}\sum_{s'=1}^{n} \partial_s(f)\partial_{s'}(g)\,\sigma(E_{is})\,\sigma(E_{i's'}),$$

and calculation yields

$$\sum_{\nu} \rho(ff_\nu\partial_i)\,\rho(g_\nu\partial_{i'}) =$$

$$\begin{cases} -f\big(\delta_{ij}\,\sigma(E_{i'j'}) + \delta_{ij'}\,\sigma(E_{i'j}) + \sigma(E_{ij})\,\sigma(E_{i'j'}) + \sigma(E_{ij'})\,\sigma(E_{i'j})\big) & \text{if } j \neq j' \\ -f\big(\delta_{ij}\,\sigma(E_{i'j}) + \sigma(E_{ij})\,\sigma(E_{i'j})\big) & \text{if } j = j'. \end{cases}$$

By the remark above we get the relations

$$\delta_{ij}\,\sigma(E_{i'j'}) + \delta_{ij'}\,\sigma(E_{i'j}) + \sigma(E_{ij})\,\sigma(E_{i'j'}) + \sigma(E_{ij'})\,\sigma(E_{i'j}) = 0,$$
$$\delta_{ij}\,\sigma(E_{i'j}) + \sigma(E_{ij})\,\sigma(E_{i'j}) = 0$$

for $1 \le i, i', j, j' \le n$ similar to those in [6] and [16]. We have written down the relations corresponding to $j' = j$ separately to make them work for $p = 2$. Note that for $p = 2$ we may assume all m_j to be > 1 since otherwise (4.3) implies $\sigma = 0$.

Lemma 4.3. *If an irreducible representation σ of the Lie algebra $\mathfrak{gl}(\mathfrak{D})$ in a vector space V satisfies the relations above then V is a coexceptional $\mathfrak{gl}(\mathfrak{D})$-module.*

For the sake of completeness we include a short proof. Let $\mathfrak{n}_+ = \langle E_{ij} \mid i > j\rangle$, $\mathfrak{h} = \langle E_{ii} \mid i = 1,\dots,n\rangle$, $\mathfrak{n}_- = \langle E_{ij} \mid i < j\rangle$. We have $\sigma(E_{ij})^2 = 0$ for $i \neq j$. Since $\mathfrak{n}_+$ acts nilpotently on V, there exists a highest weight vector $v \in V$ with respect to the Borel subalgebra $\mathfrak{h} + \mathfrak{n}_+$. Let λ be the corresponding highest weight. Put $\lambda_i = \lambda(E_{ii})$. Since $\sigma(E_{ii})^2 = -\sigma(E_{ii})$, we have $\lambda_i = 0$ or -1 for each $i = 1,\dots,n$. Furthermore, $\sigma(E_{ii})\,\sigma(E_{i'i'}) + \sigma(E_{ii'})\,\sigma(E_{i'i}) = -\sigma(E_{i'i'})$ for $i \neq i'$. Applying the endomorphisms to v, we get $\lambda_i\lambda_{i'} = -\lambda_{i'}$ for $i' > i$. Hence there exists $t = 0, 1, \dots, n$ such that $\lambda_i = -1$ for $i \le t$ and $\lambda_i = 0$ for $i > t$. As is well known, there is a unique irreducible $\mathfrak{gl}(\mathfrak{D})$-module with highest weight λ on which $\mathfrak{n}_-$ acts nilpotently. It is clear that $\bigwedge^t \mathfrak{D}^*$ is this module. The lemma is proved.

If V is an irreducible $\widehat{W}_n(k)_0$-submodule of M then V is coexceptional by Lemma 4.3. Now (4.2) yields $M = 0$.

Let further $l > 1$. Then $\sigma(\widehat{W}_n(k)_l) = 0$ and $\sigma(\widehat{W}_n(k)_{l-1})$ is a nonzero abelian ideal of the endomorphism Lie algebra $\sigma(\widehat{W}_n(k)_0)$. Define a total order relation on the set of n-tuples $A(n)$ by putting $\alpha \prec \beta$ for $\alpha, \beta \in A(n)$ if either $|\alpha| < |\beta|$ or $|\alpha| = |\beta|$ and β is greater than α in lexicographical order. For $f, g \in R$ we have

$$\begin{aligned}\rho(f\partial_i)\,\rho(g\partial_i) = fg\,\rho(\partial_i)^2 + f\partial_i(g)\,\rho(\partial_i) + \sum_{\beta>0} f\partial^\beta(g)\,\sigma(x^{(\beta)}\partial_i)\,\rho(\partial_i) \\ + \sum_{\beta>0} f\partial^{\beta+\epsilon_i}(g)\,\sigma(x^{(\beta)}\partial_i) + \sum_{\alpha>0} \partial^\alpha(f)g\,\sigma(x^{(\alpha)}\partial_i)\,\rho(\partial_i) \\ + \sum_{\alpha>0}\sum_{\beta>0} \partial^\alpha(f)\partial^\beta(g)\,\sigma(x^{(\alpha)}\partial_i)\,\sigma(x^{(\beta)}\partial_i)\,.\end{aligned}$$

Making use of the identity

$$\partial^\alpha(f)g = \sum_{\alpha'+\alpha''=\alpha} (-1)^{|\alpha''|}\binom{\alpha}{\alpha'}\partial^{\alpha'}\big(f\partial^{\alpha''}(g)\big)\,, \qquad f,\, g \in R,$$

we rewrite the last two sums occuring in the formula as

$$\begin{aligned}\sum_{\alpha=\alpha'+\alpha''>0} (-1)^{|\alpha''|}\binom{\alpha}{\alpha'}\partial^{\alpha'}\big(f\partial^{\alpha''}(g)\big)\,\sigma(x^{(\alpha)}\partial_i)\,\rho(\partial_i) \\ + \sum_{\alpha=\alpha'+\alpha''>0}\sum_{\beta>0} (-1)^{|\alpha''|}\binom{\alpha}{\alpha'}\partial^{\alpha'}\big(f\partial^{\alpha''+\beta}(g)\big)\,\sigma(x^{(\alpha)}\partial_i)\,\sigma(x^{(\beta)}\partial_i)\,.\end{aligned}$$

Note that the nonzero terms occur for $|\alpha| \le l$, $|\beta| \le l$ only.

Consider the case $p = 0$. Let $\gamma \in A(n)$. Applying Proposition 3.4, we find a finite number of elements f_ν, $g_\nu \in R$ such that

$$\sum_\nu f_\nu \partial^\alpha(g_\nu) = \begin{cases} 0 & \text{whenever } |\alpha| \le 2l,\ \alpha \ne 2\gamma \\ 1 & \text{for } \alpha = 2\gamma\,. \end{cases}$$

We get

$$\sum_\nu \rho(ff_\nu\partial_i)\,\rho(g_\nu\partial_i) = f \sum_{\alpha+\beta=2\gamma} (-1)^l\,\sigma(x^{(\alpha)}\partial_i)\,\sigma(x^{(\beta)}\partial_i)$$

for all $f \in R$. It follows $\sum_{\alpha+\beta=2\gamma} \sigma(x^{(\alpha)}\partial_i)\,\sigma(x^{(\beta)}\partial_i) = 0$, i. e.

$$\sigma(x^{(\gamma)}\partial_i)^2 = -2 \sum_{\substack{\alpha+\beta=2\gamma,\\ \alpha>\gamma}} \sigma(x^{(\alpha)}\partial_i)\,\sigma(x^{(\beta)}\partial_i)\,.$$

Note that the endomorphisms $\sigma(x^{(\alpha)}\partial_i)$, $|\alpha| = l$, are pairwise commuting. Assuming that the endomorphisms $\sigma(x^{(\alpha)}\partial_i)$ are nilpotent for all $\alpha \in A(n)$ such that $|\alpha| = l$, $\alpha \succ \gamma$, we deduce that $\sigma(x^{(\gamma)}\partial_i)$ is nilpotent as well. Hence induction ensures that $\sigma(x^{(\alpha)}\partial_i)$ are nilpotent for all α with $|\alpha| = l$. We see that $\sigma(\widehat{W}_n(k)_{l-1})$ consists of nilpotent endomorphisms. For any irreducible $\widehat{W}_n(k)_0$-submodule V of M the subspace $V' = \{v \in V \mid \sigma(\widehat{W}_n(k)_{l-1})\,v = 0\}$ is a nonzero $\widehat{W}_n(k)_0$-submodule of V, whence $V' = V$, i. e. $\sigma(\widehat{W}_n(k)_{l-1})$ vanishes on V. The complete reducibility of M now implies that $\sigma(\widehat{W}_n(k)_{l-1}) = 0$, which contradicts the definition of l.

The same arguments work for $p > 0$ provided that $\sigma(x^{(\alpha)}\partial_i) = 0$ for all α and i whenever $\alpha_j \ge p^{m_j}/2$ for at least one j. To achieve the full generality in positive characteristic we need to work harder.

Lemma 4.4. *Let a, m be positive integers with $a < p^m$. Then there exists a positive integer b such that $\binom{a}{b} \not\equiv 0 \pmod p$ and $a + b < p^m$ unless $b = p^m - p^r$ for some r, $0 \le r < m$.*

Proof. Consider the p-adic expansion $a = a_0 + a_1 p + \cdots$, where $0 \le a_i < p$. Let r be the least integer such that $a_r \neq 0$. Then $b = p^r$ is the least positive integer for which p does not divide the binomial coefficient. We have $a + b < p^m$ unless $a = p^m - p^r$.

For a family $\Phi = (\varphi_\alpha)_{\alpha \in A(n)}$ of R-linear endomorphisms of M we define its support $\operatorname{supp} \Phi = \{\alpha \in A(n) \mid \varphi_\alpha \neq 0\}$. In case of a finite support define the degree $\deg \Phi$ of Φ to be the largest integer t such that there exists $\alpha \in \operatorname{supp} \Phi$ with $|\alpha| = t$.

Lemma 4.5. *Let $t \ge 0$ be an integer and Φ a family of R-linear endomorphisms satisfying*

(4.4) $\operatorname{supp} \Phi$ *is a finite subset of* $A'(n : \mathbf{m})$ *and* $\deg \Phi \le t$,

(4.5) *the endomorphisms* φ_α *with* $|\alpha| = t$ *are pairwise commuting,*

(4.6) M' *is stable under all endomorphisms* $\Phi(f) = \sum_{\alpha \in A(n)} \partial^\alpha(f) \varphi_\alpha$, $f \in R$.

Then the endomorphisms φ_α *are nilpotent for all* α *with* $|\alpha| = t$.

Proof. If $t = 0$ then $\Phi(f) = f\varphi_0$, and the hypothesis that M' contains no nonzero R-submodules ensures $\varphi_0 = 0$. Assume that $t > 0$ and the lemma is valid for the less values of t. Let $\gamma \in A'(n : \mathbf{m})$, $|\gamma| = t$. Assume that φ_α are nilpotent for all α with $|\alpha| = t$, $\alpha \succ \gamma$. Under these assumptions we shall prove that φ_γ is nilpotent as well. The induction then will complete the proof.

Choose $\gamma' \in A(n)$, $\gamma' > 0$, such that $\binom{\gamma}{\gamma'} \not\equiv 0 \pmod{p}$ and $\gamma + \gamma' \in A(n : \mathbf{m})$. This is possible by Lemma 4.4. We can take $\gamma' = b\epsilon_j$ for some j such that $\gamma_j > 0$, and b then must satisfy $\binom{\gamma_j}{b} \not\equiv 0 \pmod{p}$, $\gamma_j + b < p^{m_j}$. Put $t' = |\gamma'|$. By Proposition 3.5 there exists a finite number of elements $f_\nu, g_\nu \in R$ such that

$$\sum_\nu f_\nu \partial^\alpha(g_\nu) = \begin{cases} 0 & \text{whenever } \alpha \in A(n : \mathbf{m}),\ |\alpha| \le 2t,\ \alpha \neq \gamma + \gamma' \\ 1 & \text{for } \alpha = \gamma + \gamma'. \end{cases}$$

Consider the endomorphisms

$$\begin{aligned} \sum_\nu \Phi(f f_\nu) \Phi(g_\nu) &= \sum_\alpha \sum_\beta \sum_\nu \partial^\alpha(f f_\nu) \partial^\beta(g_\nu) \varphi_\alpha \varphi_\beta \\ &= \sum_{\alpha' + \alpha'' = \alpha} \sum_\beta \sum_\nu (-1)^{|\alpha''|} \binom{\alpha}{\alpha'} \partial^{\alpha'}\big(f f_\nu \partial^{\alpha'' + \beta}(g_\nu)\big) \varphi_\alpha \varphi_\beta \end{aligned}$$

with $f \in R$. Denote by $S_{\alpha'\alpha''\beta}$ the sum over ν of the terms with fixed values of α', α'', β. In examining those partial sums we may assume $|\alpha| \le t$ and $|\beta| \le t$, since otherwise all terms trivially vanish. If $\alpha'' + \beta \in A(n : \mathbf{m})$ then $S_{\alpha'\alpha''\beta} = 0$ unless $\alpha'' + \beta = \gamma + \gamma'$. If $\alpha'' + \beta = \gamma + \gamma'$ then $S_{\alpha'\alpha''\beta} = (-1)^{|\alpha''|} \binom{\alpha}{\alpha'} \partial^{\alpha'}(f) \varphi_\alpha \varphi_\beta$. In this case $|\alpha''| \ge t'$ since $|\beta| \le t$, and $|\alpha'| \le t - t'$ since $|\alpha| \le t$. Note that $|\alpha'| = t - t'$ implies $|\alpha''| = t'$, $|\alpha| = |\beta| = t$. Consider the case $\alpha'' + \beta \notin A(n : \mathbf{m})$. The assumptions on $\partial_1, \ldots, \partial_n$ ensure that $\partial^{\alpha'' + \beta}$ is a linear combination with scalar coefficients of the differential operators ∂^η with $\eta \in A(n : \mathbf{m})$, $|\eta| < |\alpha'' + \beta| = |\alpha''| + |\beta|$. If $\partial^{\gamma + \gamma'}$ does not occur in this linear combination then $S_{\alpha'\alpha''\beta} = 0$.

If $\partial^{\gamma+\gamma'}$ does occur then $S_{\alpha'\alpha''\beta}$ is a scalar multiple of $(-1)^{|\alpha''|}\binom{\alpha}{\alpha'}\partial^{\alpha'}(f)\varphi_\alpha\varphi_\beta$. In the latter case $t+t' = |\gamma+\gamma'| < |\alpha''|+|\beta|$, whence $|\alpha''| > t'$, and $|\alpha'| < t-t'$. Now we have

$$\sum_\nu \Phi(ff_\nu)\,\Phi(g_\nu) = \sum_{\alpha'} \partial^{\alpha'}(f)\psi_{\alpha'} = \Psi(f),$$

where $\Psi = (\psi_{\alpha'})_{\alpha'\in A(n)}$ is a collection of R-linear endomorphisms of M. Obviously, M' is stable under $\Psi(f)$. Next, $\operatorname{supp}\Psi \subseteq \{\alpha' \in A(n) \mid \alpha' \leq \alpha$ for some $\alpha \in \operatorname{supp}\Phi\}$, and so (4.4) is satisfied for the family Ψ. The analysis above shows also that $\deg\Psi \leq t-t'$. Furthermore,

$$\psi_{\alpha'} = \sum (-1)^{t'}\binom{\alpha}{\alpha'}\varphi_\alpha\varphi_\beta \qquad \text{for} \quad |\alpha'| = t-t',$$

where the sum is taken over all α, $\beta \in \operatorname{supp}\Phi$ such that $|\alpha| = |\beta| = t$, $\alpha \geq \alpha'$, $\alpha - \alpha' + \beta = \gamma + \gamma'$. We see that all endomorphisms $\psi_{\alpha'}$ with $|\alpha'| = t - t'$ are contained in the commutative subalgebra generated by the endomorphisms φ_α with $|\alpha| = t$. Applying the induction hypotheses, we conclude that $\psi_{\alpha'}$ are nilpotent for all α' such that $|\alpha'| = t-t'$. In particular, $\psi_{\gamma-\gamma'}$ is nilpotent. Now

$$\binom{\gamma}{\gamma'}\varphi_\gamma^2 = (-1)^{t'}\psi_{\gamma-\gamma'} - \sum \binom{\alpha}{\gamma-\gamma'}\varphi_\alpha\varphi_\beta,$$

where the sum is taken over all α, $\beta \in \operatorname{supp}\Phi$ such that $|\alpha| = |\beta| = t$, $\alpha \geq \gamma-\gamma'$, $\alpha+\beta = 2\gamma$, $\alpha \neq \gamma$. For each term of the sum either $\alpha \succ \gamma$ or $\beta \succ \gamma$. By the hypotheses all terms are nilpotent. Hence φ_γ is nilpotent as well. The lemma is proved.

For a fixed i put $\varphi_0 = 0$, $\varphi_\alpha = \sigma(x^{(\alpha)}\partial_i)$ for $\alpha > 0$, and $t = l$. The properties (4.4), (4.5) then fulfill but (4.6) is lacking, so that we can not apply Lemma 4.5 directly. However, M' is stable under the endomorphisms $\rho(f\partial_i) = f\,\rho(\partial_i)+\Phi(f)$. If $\gamma' \in A(n)$ and f_ν, $g_\nu \in R$ are chosen as in Lemma 4.5, we see that

$$\sum_\nu \rho(ff_\nu\partial_i)\,\rho(g_\nu\partial_i) = \sum_\nu \Phi(ff_\nu)\,\Phi(g_\nu) = \Psi(f).$$

Hence M' is stable under the endomorphisms $\Psi(f)$. The arguments used in the proof of the lemma now work without changes. We deduce that $\sigma(x^{(\alpha)}\partial_i)$ are nilpotent for all $\alpha \in A(n)$ with $|\alpha| = l$ and $i = 1,\dots,n$. It follows from the assumption of complete reducibility that $\sigma(\widehat{W}_n(k)_{l-1}) = 0$, a contradiction. The proof of Theorem 4.2 is completed.

We consider now the modules $\mathcal{R}V$ and, more generally, the modules $P\otimes_R\mathcal{R}V$, where P is an object of $\mathcal{C}$ with $\sigma_P = 0$. Let $\widetilde{V}$ be a second discrete $\widehat{W}_n(k)_0$-module and $\widetilde{P}$ a second object of $\mathcal{C}$ with $\sigma_{\widetilde{P}} = 0$. Put

$$M = P\otimes_R\mathcal{R}V \cong P\otimes_k V, \qquad \widetilde{M} = \widetilde{P}\otimes_R\mathcal{R}\widetilde{V} \cong \widetilde{P}\otimes_k\widetilde{V}.$$

There are two cases where the description of $\mathcal{C}$-submodules and morphisms in $\mathcal{C}$ is especially simple.

Proposition 4.6. *Assume that* $V, \widetilde{V}$ *are irreducible discrete* $\mathfrak{g}$*-modules and* $\mathrm{End}_{\mathfrak{g}} V \cong k, \mathrm{End}_{\mathfrak{g}} \widetilde{V} \cong k$, *where* $\mathfrak{g} = \widehat{W}_n(k)_0$. *Then*

(i) *The assignment* $P' \mapsto P' \otimes_R \mathcal{R}V \cong P' \otimes_k V$ *establishes a one-to-one correspondence between the* $(R, \mathfrak{D})$*-submodules of* P *and the* $\mathcal{C}$*-submodules of* M.

(ii) $\mathrm{Mor}_{\mathcal{C}}(M, \widetilde{M}) = 0$ *when* $V \not\cong \widetilde{V}$, *and* $\mathrm{Mor}_{\mathcal{C}}(M, \widetilde{M}) \cong \mathrm{Hom}_{(R,\mathfrak{D})}(P, \widetilde{P})$ *when* $V \cong \widetilde{V}$.

Proof. The class of $(R, \mathfrak{D})$-submodules of P and that of $\mathcal{C}$-submodules coincide. Hence $P' \otimes_R \mathcal{R}V$ is a $\mathcal{C}$-submodule of M for any $(R, \mathfrak{D})$-submodule P' of P. For each $q \in P$ the mapping $\xi_q : V \to M$, $v \in V \mapsto q \otimes v$, is a $\mathfrak{g}$-module homomorphism. Hence $M = \sum_{q \in P} \xi_q(V)$ is completely reducible as a $\mathfrak{g}$-module and all its irreducible submodules are isomorphic with V. The assumption $\mathrm{End}_{\mathfrak{g}} V \cong k$ implies that any $\mathfrak{g}$-module homomorphism $V \to M$ is of the form ξ_q for some $q \in P$. In other words, $\mathrm{Hom}_{\mathfrak{g}}(V, M) \cong P$. The R-module and $\mathfrak{D}$-module structures on P are expressed in terms of this isomorphism as follows: $(f\xi)(v) = f \cdot \xi(v)$ and $(D\xi)(v) = D(\xi(v))$ for $f \in R$, $D \in \mathfrak{D}$, $\xi \in \mathrm{Hom}_{\mathfrak{g}}(V, M)$, $v \in V$. Given a $\mathcal{C}$-submodule M' of M, we see that $\mathrm{Hom}_{\mathfrak{g}}(V, M')$ is an $(R, \mathfrak{D})$-submodule of $\mathrm{Hom}_{\mathfrak{g}}(V, M)$, which corresponds therefore to a certain $(R, \mathfrak{D})$-submodule P' of P. Now M', being a $\mathfrak{g}$-submodule of M, is a sum of its irreducible $\mathfrak{g}$-submodules. Hence $M' = \sum_{q \in P'} \xi_q(V) = P' \otimes_k V$. It follows that the assignment $M' \mapsto \mathrm{Hom}_{\mathfrak{g}}(V, M')$ defines a mapping inverse to the one indicated in the statement of (i). That completes the proof of (i).

As for (ii), notice that $\widetilde{M}$ is a sum of its irreducible $\mathfrak{g}$-submodules isomorphic to $\widetilde{V}$. Hence $\mathrm{Hom}_{\mathfrak{g}}(M, \widetilde{M}) = 0$ unless $V \cong \widetilde{V}$. Assume $V = \widetilde{V}$. If $\varphi : M \to \widetilde{M}$ is a morphism in $\mathcal{C}$ then its graph $\Gamma(\varphi)$ is a $\mathcal{C}$-submodule of $M \oplus \widetilde{M}$ such that the projection $M \oplus \widetilde{M} \to M$ induces an isomorphism $\Gamma(\varphi) \xrightarrow{\sim} M$. By (i) $\Gamma(\varphi) = P' \otimes_R \mathcal{R}V$, where P' is an $(R, \mathfrak{D})$-submodule of $P \oplus \widetilde{P}$. It is clear that the projection $P \oplus \widetilde{P} \to P$ induces an isomorphism $P' \xrightarrow{\sim} P$. Hence $P' = \Gamma(\psi)$ for an $(R, \mathfrak{D})$-module homomorphism $\psi : P \to \widetilde{P}$. We conclude $\varphi = \psi \otimes \mathrm{id}$.

Proposition 4.7. *Assume that* $P, \widetilde{P}$ *are simple objects of* $\mathcal{C}$ *and* $\mathrm{End}_{\mathcal{C}} P \cong k$, $\mathrm{End}_{\mathcal{C}} \widetilde{P} \cong k$ *(these assumptions hold, in particular, whenever* R *is* $\mathfrak{D}$*-simple,* $R^{\mathfrak{D}} \cong k$ *and* $P, \widetilde{P}$ *are invertible as* R*-modules). Put* $\mathfrak{g} = \widehat{W}_n(k)_0$. *Then*

(i) *The assignment* $V' \mapsto P \otimes_R \mathcal{R}V' \cong P \otimes_k V'$ *establishes a one-to-one correspondence between the* $\mathfrak{g}$*-submodules of* V *and the* $\mathcal{C}$*-submodules of* M.

(ii) $\mathrm{Mor}_{\mathcal{C}}(M, \widetilde{M}) = 0$ *if* $P \not\cong \widetilde{P}$ *and* $\mathrm{Mor}_{\mathcal{C}}(M, \widetilde{M}) \cong \mathrm{Hom}_{\mathfrak{g}}(V, \widetilde{V})$ *if* $P \cong \widetilde{P}$.

Proof. The hypotheses mean that P is an irreducible $(R, \mathfrak{D})$-module and $\mathrm{End}_{(R,\mathfrak{D})} P \cong k$. We see that M is a sum of its irreducible $(R, \mathfrak{D})$-submodules isomorphic to P and $V \cong \mathrm{Hom}_{(R,\mathfrak{D})}(P, M)$. As compared with the proof of Proposition 4.6, P and V interchanged their roles. Otherwise the arguments are the same. Similarly, (ii) is a consequence of (i).

Theorem 4.8. *Let $V, \widetilde{V}$ be irreducible discrete $\widehat{W}_n(k)_0$-modules distinct from the coexceptional height one modules. If $p > 0$, assume in addition that $V, \widetilde{V}$ are annihilated by $x^{(\alpha)}\partial_i$ whenever $\alpha \notin A'(n : \mathbf{m})$. Let $P, \widetilde{P}$ be simple objects of $\mathcal{C}$ with $\sigma_P = 0$, $\sigma_{\widetilde{P}} = 0$ and $\operatorname{End}_R P \cong k$, $\operatorname{End}_R \widetilde{P} \cong k$. Then $P \otimes_R \mathcal{R}V$ and $\widetilde{P} \otimes_R \mathcal{R}\widetilde{V}$ are irreducible $R\mathfrak{D}$-modules. In order that these modules be isomorphic, it is necessary and sufficient that $P \cong \widetilde{P}$ and $V \cong \widetilde{V}$.*

This is a consequence of Theorem 4.2 and Proposition 4.7. Combining Theorem 4.8 with Theorems 2.2, 2.3, we get

Theorem 4.9. *Let $V, \widetilde{V}$ be irreducible discrete finite dimensional $\widehat{W}_n(k)_0$-modules distinct from the coexceptional height one modules. Then the $\widehat{W}_n(k)$-modules* $\operatorname{Coind} V$ *and* $\operatorname{Coind} \widetilde{V}$ *coinduced from V and $\widetilde{V}$ are irreducible. In order that they be isomorphic, it is necessary and sufficient that $V \cong \widetilde{V}$.*

Theorem 4.10. *Let $V, \widetilde{V}$ be irreducible discrete $\widehat{W}_n(k)_0$-modules distinct from the exceptional height one modules. Then the $\widehat{W}_n(k)$-modules* $\operatorname{Ind} V$ *and* $\operatorname{Ind} \widetilde{V}$ *induced from V and $\widetilde{V}$ are irreducible. In order that they be isomorphic, it is necessary and sufficient that $V \cong \widetilde{V}$.*

Note that $V \cong \bigwedge^t \mathfrak{D}$ for some $t = 0, \ldots, n$ if and only if $V \otimes_k \bigwedge^n \mathfrak{D}^* \cong \bigwedge^{n-t} \mathfrak{D}^*$.

Remarks. In case of characteristic zero Theorem 4.10 is due to Rudakov [19]. Our technique handles the induced and the coinduced modules equally well.

The assumption of complete reducibility in the statement of Theorem 4.2 can not be dropped. Let $R = \widehat{O}_n(k)$. Consider the $\widehat{W}_n(k)_0$-modules

$$V'' = \widehat{W}_n(k)_0/\widehat{W}_n(k)_1, \qquad V = \widehat{W}_n(k)/\widehat{W}_n(k)_1, \qquad V' = \widehat{W}_n(k)/\widehat{W}_n(k)_0.$$

The exact sequence of $\widehat{W}_n(k)_0$-modules

$$0 \to V'' \to V \to V' \to 0 \tag{$*$}$$

gives rise to an exact sequence in $\mathcal{C}$:

$$0 \to \operatorname{Coind} V'' \to \operatorname{Coind} V \to \operatorname{Coind} V' \to 0. \tag{$**$}$$

We know that $\operatorname{Coind} V'$ is isomorphic to the adjoint $\widehat{W}_n(k)$-module. The projection $\widehat{W}_n(k) \to V$ is a homomorphism of $\widehat{W}_n(k)_0$-modules. There is the induced homomorphism of $\widehat{W}_n(k)$-modules $\operatorname{Coind} V' \cong \widehat{W}_n(k) \to \operatorname{Coind} V$, which is easily seen to be a splitting for $(**)$. However, $(**)$ does not split in $\mathcal{C}$ because any such a splitting would correspond, by Proposition 4.7, to a splitting of $(*)$, which has none. Thus $\operatorname{Coind} V$ contains a $\widehat{W}_n(k)$-submodule complementary to $\operatorname{Coind} V''$, which is not a $\mathcal{C}$-submodule.

5. Representations of generalized Witt algebras

Let k be a field of characteristic $p > 0$. To an n-tuple $\mathbf{m} = (m_1, \dots, m_n)$ of positive integers there corresponds a $\mathfrak{D}$-invariant subalgebra $O_n(\mathbf{m}; k)$ of $O_n(k)$ whose basis consists of the monomials $x^{(\alpha)}$ with $\alpha \in A(n : \mathbf{m})$ [15]. The Lie algebra $W_n(\mathbf{m}; k) = O_n(\mathbf{m}; k)\mathfrak{D}$ of its special derivations is called a generalized Witt algebra. Put $W_n(\mathbf{m}; k)_0 = W_n(\mathbf{m}; k) \cap \widehat{W}_n(k)_0$.

We note that the construction of modules in Section 1 makes sense in slightly modified settings. Suppose that $\partial_i^{p^{m_i}} = 0$ as a derivation of the algebra R for each $i = 1, \dots, n$. Then we may assume V to be a $W_n(\mathbf{m}; k)_0$-module and thus view $\mathfrak{R}$ as a functor from the category of $W_n(\mathbf{m}; k)_0$-modules to the category of $R\mathfrak{D}$-modules or, more precisely, to the category of compatible structures of an $R\mathfrak{D}$-module, an R-module and a $W_n(\mathbf{m}; k)_0$-module. Here is an explanation from the viewpoint of Section 2. For each algebra homomorphism $\xi : R \to K$ the corresponding Lie algebra homomorphism $j_\xi : R\mathfrak{D} \to \widehat{W}_n(K)$ has image in $W_n(\mathbf{m}; K)$. Hence $R\mathfrak{D}$ operates in the $W_n(\mathbf{m}; K)$-module $\operatorname{Coind} V_K$ coinduced from the $W_n(\mathbf{m}; K)_0$-module V_K. There is a canonical isomorphism $\mathfrak{J}_\xi(\mathfrak{R}V) \cong \operatorname{Coind} V_K$. A verification similar to that in Section 2 shows that for the mapping $j_\xi : \mathfrak{J}_\xi(\mathfrak{R}V) \to \operatorname{Coind} V_K$ we have $j_\xi(\rho(D)m) = j_\xi(D) \cdot j_\xi(m)$ for all $D \in R\mathfrak{D}$, $m \in \mathfrak{R}V$. Thus $\mathfrak{R}V$ is identified with an $R\mathfrak{D}$-submodule of $\operatorname{Coind} V_R$. In fact, if the representation σ_V of $W_n(\mathbf{m}; k)_0$ in V satisfies $\sigma_V(x^{(\alpha)}\partial_i) = 0$ for all $\alpha \in A(n : \mathbf{m})$ such that $\alpha_j = p^{m_j - 1}$ for at least one j, then we can extend σ_V to a representation of $\widehat{W}_n(k)_0$ by putting $\sigma_V(x^{(\alpha)}\partial_i) = 0$ for all $\alpha \in A(n) \setminus A(n : \mathbf{m})$. Then we find ourselves in the old settings.

Let further $R = O_n(\mathbf{m}; k)$. Then $\partial_i^{p^{m_i}} = 0$ and $\{\partial_i^{p^r} \mid 1 \le i \le n,\ 0 \le r < m_i\}$ is a system of linearly independent over k derivations of R. It is independent by Proposition 3.1 because the algebra R is $\mathfrak{D}$-simple. Now $\mathfrak{R}$ is a functor from the category of $W_n(\mathbf{m}; k)_0$-modules to that of $W_n(\mathbf{m}; k)$-modules, and we meet the general assumptions of Section 4.

Similarly to Theorem 2.2 we can describe the modules we consider now as truncated coinduced modules. Let $L = W_n(\mathbf{m}; k)$, or more generally let L be an arbitrary transitive subalgebra of $W_n(\mathbf{m}; k)$. Let $P(L)$ be the universal p-envelope of L, i. e. the p-Lie subalgebra in the universal enveloping algebra $U(L)$ generated by L. Denote by $\mathfrak{m}$ the maximal ideal of R, and by $\mathfrak{h}$ its normalizer in $P(L)$. Then $\mathfrak{h}$ is a p-subalgebra of $P(L)$. Its restricted universal enveloping algebra $u(\mathfrak{h})$ is identified with a subalgebra of $u(P(L)) \cong U(L)$. For a restricted $\mathfrak{h}$-module V the L-modules induced and coinduced from V are defined as

$$\operatorname{ind} V = U(L) \otimes_{u(\mathfrak{h})} V \qquad \text{and} \qquad \operatorname{coind} V = \operatorname{Hom}_{u(\mathfrak{h})}(U(L), V)$$

with the obvious L-module structures.

Theorem 5.1. (i) *If M is an (R, L)-module then* $M \cong \operatorname{coind}(M/\mathfrak{m}M)$.

(ii) $\operatorname{ind} V \cong \operatorname{coind}\big(V \otimes \bigwedge^n (L/L_0)^*\big)$.

Part (i) is due to Kuznetsov [17, Th. 2.1] (see also [21, Th. 2.3, Th. 4.2]).

Assertions similar to part (ii) of the theorem are obtained independently by Dzhumadil'daev [5, §5], Farnsteiner and Strade [8], Kuznetsov [17, Th. 2.1]. In the particular case $L = W_n(\mathbf{m}; k)$ it was first announced in [20].

Let further $L = W_n(\mathbf{m}; k)$. Then $\mathfrak{h}$ is the sum of $P(L_0)$ and the p-subalgebra generated by the central elements $z_i = \partial_i^{p^{m_i}}$, $i = 1, \dots, n$. Given an n-tuple $\lambda = (\lambda_1, \dots, \lambda_n) \in k^n$, we can make each L_0-module V into a restricted $\mathfrak{h}$-module by letting each z_i operate on V as a scalar multiplication by λ_i. Denote by $\operatorname{ind}_\lambda V$ and $\operatorname{coind}_\lambda V$ respectively the L-modules induced and coinduced from the $\mathfrak{h}$-module thus obtained. Denoting by I_λ the ideal of $U(L)$ generated by $z_1 - \lambda_1, \dots, z_n - \lambda_n$, we have

$$\operatorname{ind}_\lambda V \cong U(L)/I_\lambda \otimes_{U(L_0)} V, \qquad \operatorname{coind}_\lambda V \cong \operatorname{Hom}_{U(L_0)}(U(L)/I_\lambda, V).$$

To each $\lambda \in k^n$ we put into correspondence a closed differential 1-form $\psi_\lambda = \sum_{i=1}^n \lambda_i x_i^{(p^{m_i}-1)} dx_i$. As first noted by Kryluk [16], each (R, L)-module M admits a twisting of its L-module structure by ψ_λ. The twisting results in a new (R, L)-module which we can describe as the tensor product $P_\lambda \otimes_R M$, where P_λ is the (R, L)-module obtained by twisting the natural (R, L)-module R. Thus P_λ has R as its underlying R-module, and the action ρ_λ of L is given by the rule $\rho_\lambda(D)g = Dg + \psi_\lambda(D)\, g$ for $g \in R$, $D \in L$. Jacobson's p-th power formula shows that $\rho_\lambda(\partial_i)^{p^r} = \partial_i^{p^r} + \lambda_i x_i^{(p^{m_i}-p^r)}$ for $r = 0, \dots, m_i$. Hence the p^{m_i}-th power of $\rho_\lambda(\partial_i)$ is a scalar multiplication by λ_i. We may view the P_λ's as objects of the category $\mathcal{C}$ introduced in Section 1. These are precisely the invertible objects P with $\sigma_P = 0$: by [21, Th. 7.1, Th. 7.4] any such a P is isomorphic to exactly one P_λ. In particular, the P_λ's are simple objects of $\mathcal{C}$.

Proposition 5.2. *Let V be a $W_n(\mathbf{m}; k)_0$-module and $\lambda \in k^n$. Then* $\operatorname{coind}_\lambda V \cong P_\lambda \otimes_R \mathcal{R}V$ *and* $\operatorname{ind}_\lambda V \cong P_\lambda \otimes_R \mathcal{R}(V \otimes_k \bigwedge^n (L/L_0)^*)$.

Proof. Put $M = P_\lambda \otimes_R \mathcal{R}V$. Since $\rho_\lambda(fD)\, g = f \cdot \rho_\lambda(D)g \in \mathfrak{m}$ for $f \in \mathfrak{m}$, $D \in L$, $g \in R$, the Lie algebra L_0 operates trivially in $P_\lambda/\mathfrak{m}P_\lambda$. As to the representation ρ in $\mathcal{R}V$, we have

$$\begin{aligned}\rho(x^{(\beta)}\partial_i)\,(g \otimes v) &= x^{(\beta)}\partial_i(g) \otimes v + \sum_{0<\alpha\le\beta} x^{(\beta-\alpha)} g \otimes \sigma_V(x^{(\alpha)}\partial_i)\, v \\ &\equiv g \otimes \sigma_V(x^{(\beta)}\partial_i)\, v \pmod{\mathfrak{m}(\mathcal{R}V)}\end{aligned}$$

for $\beta \in A(n : \mathbf{m})$, $\beta > 0$, $g \in R$, $v \in V$, whence the canonical linear isomorphism $\mathcal{R}V/\mathfrak{m}(\mathcal{R}V) \cong R/\mathfrak{m} \otimes_k V \cong V$ is an isomorphism of L_0-modules. Hence so is the isomorphism $M/\mathfrak{m}M \cong P_\lambda/\mathfrak{m}P_\lambda \otimes_k \mathcal{R}V/\mathfrak{m}(\mathcal{R}V) \cong V$ as well. Now $\partial_i^{p^{m_i}}$ annihilates $\mathcal{R}V$. It operates as a scalar multiplication by λ_i in P_λ, hence also in M and $M/\mathfrak{m}M$. It remains to apply Theorem 5.1.

We can state now a special case of Theorem 4.8.

Theorem 5.3. (i) *Let V be an (absolutely) irreducible L_0-module, and let $\lambda \in k^n$. Assume that V is annihilated by $x^{(\alpha)}\partial_i$ whenever $\alpha \in A(n : \mathbf{m}) \setminus A'(n : \mathbf{m})$. The truncated coinduced L-module* $\operatorname{coind}_\lambda V$ *is (absolutely) irreducible unless V is* *coexceptional. The truncated induced L-module* $\mathrm{ind}_\lambda V$ *is (absolutely) irreducible unless* V *is exceptional.*

(ii) *Let* V, $\widetilde{V}$ *be irreducible* L_0*-modules, and let* λ, $\widetilde{\lambda} \in k^n$. *Assume that* V, $\widetilde{V}$ *are annihilated by* $x^{(\alpha)}\partial_i$ *whenever* $\alpha \in A(n:\mathbf{m}) \setminus A'(n:\mathbf{m})$. *Under assumption that neither* V, *nor* $\widetilde{V}$ *are coexceptional,* $\mathrm{coind}_\lambda V \cong \mathrm{coind}_{\widetilde{\lambda}} \widetilde{V}$ *if and only if* $V \cong \widetilde{V}$ *and* $\lambda = \widetilde{\lambda}$. *Under assumption that neither* V, *nor* $\widetilde{V}$ *are exceptional,* $\mathrm{ind}_\lambda V \cong \mathrm{ind}_{\widetilde{\lambda}} \widetilde{V}$ *if and only if* $V \cong \widetilde{V}$ *and* $\lambda = \widetilde{\lambda}$.

Remarks. The irreducible representations of the algebras $W_n(\mathbf{m}; k)$ were studied by Chang [2], Strade [23] ($n = 1$, $m = 1$), Mil'ner [18] ($n = 1$), Dzhumadil'daev [3] (representations of the maximal subalgebras $W_1(m;k)_0$), Koreshkov [13], [14] ($\mathbf{m} = \mathbf{1}$), Ermolaev [6], Kryluk [16] (height one representations). As is well known, any absolutely irreducible $W_1(m;k)$-module containing an irreducible $W_1(m;k)_0$-module V of maximal height coincides with V. In particular, the absolute irreducibility of induced and coinduced modules fail for such a V. That accounts for the imposed restrictions on the n-tuples $\alpha \in A(n:\mathbf{m})$ with nonzero endomorphisms $\sigma_V(x^{(\alpha)}\partial_i)$. It seems that $p^{m_i} - p^{m_i - 1}$, $i = 1, \dots, n$, are the best upper bounds for the components of such n-tuples, at least in case $n = 1$. Although stated to the contrary in [18] and [4, §3], there appears to be a class of irreducible $W_1(m;k)_0$-modules of height $p^m - p^{m-1} \le l < p^m - 1$ with reducible induced and coinduced modules. As an illustration we consider the simplest example:

Example. Let $p = 3$, $n = 1$, $m = 2$. In the proof of Theorem 4.2 we were dealing with the situation when M' is an $R\mathfrak{D}$-submodule of M containing no nonzero R-submodules and satisfying $RM' = M$. We shall use the short notation σ_r for the endomorphisms $\sigma_M(x^{(r+1)}\partial)$. The technique we used first breaks down when $\sigma_7 = \sigma_6 = 0$, $\sigma_5 \neq 0$. Let f_ν, $g_\nu \in R$ be a finite number of elements such that

$$\sum_\nu f_\nu \partial^\alpha(g_\nu) = \begin{cases} 0 & \text{for } \alpha = 0, 1, \dots, 7 \\ 1 & \text{for } \alpha = 8. \end{cases}$$

M' is stable under the endomorphisms

$$\sum_\nu \rho_M(ff_\nu\partial)\,\rho_M(g_\nu\partial) = -\partial^3(f)\sigma_5\sigma_4 - \partial^2(f)\sigma_4^2 + \partial(f)\sigma_4\sigma_3 + f(-\sigma_5\sigma_1 + \sigma_4\sigma_2 + \sigma_3^2)$$

with $f \in R$. By Lemma 4.5 $\sigma_5\sigma_4$ is nilpotent. For any irreducible $\sigma(L_0)$-submodule $V \subseteq M$ the subspace $\{v \in V \mid \sigma_5\sigma_4 v = 0\}$ is nonzero. As $[\sigma_r, \sigma_5\sigma_4] = 0$ for all $r = 0, \dots, 7$, it is a $\sigma(L_0)$-submodule, hence coincides with V. We conclude $\sigma_5\sigma_4 = 0$. Repeating twice the same procedure we get $\sigma_4 = 0$ and $-\sigma_5\sigma_1 + \sigma_3^2 = 0$. Thus the conclusion of the theorem remains valid unless M contains irreducible $\sigma(L_0)$-submodules satisfying

$$\sigma_7 = \sigma_6 = 0, \quad \sigma_5 \neq 0, \quad \sigma_4 = 0, \quad \sigma_5\sigma_1 - \sigma_3^2 = 0. \qquad (*)$$

Similar, though longer, computations show that for $\sigma_7 = 0$, $\sigma_6 \neq 0$ we encounter irreducible $\sigma(L_0)$-modules satisfying

$$\begin{aligned}&\sigma_7 = 0\,, \quad \sigma_6 \neq 0\,, \quad \sigma_6\sigma_3 - \sigma_5\sigma_4 = 0\,,\\ &\sigma_6^2\sigma_0 - \sigma_6\sigma_5\sigma_1 + \sigma_6\sigma_4\sigma_2 + \sigma_6\sigma_3^2 + \sigma_6^2 + \sigma_4^3 = 0\,.\end{aligned} \tag{$**$}$$

Now, if M does satisfy either $(*)$ or $(**)$, then there exists a nontrivial $R\mathfrak{D}$-module endomorphism of M having the form

$$Z = \xi\,\rho_M(\partial)^3 + \eta\,\rho_M(\partial) + \zeta\,,$$

where endomorphisms ξ, η, ζ belong to the associative subalgebra generated by the endomorphisms σ_r, $r = 0, \dots, 7$. As

$$\begin{aligned}[\rho_M(f\partial),\, Z] &= [f\,\rho_M(\partial) + \sum_{r=0}^{7} (\partial^{r+1} f)\,\sigma_r,\, Z]\\ &= -\big((\partial^3 f)\,\xi + (\partial f)\,\eta\big)\,\rho_M(\partial) + \sum_{r=0}^{7} (\partial^{r+1} f)\,[\sigma_r,\, \xi]\,\rho_M(\partial)^3\\ &\quad + \sum_{r=0}^{7} (\partial^{r+1} f)\,[\sigma_r,\, \eta]\,\rho_M(\partial)\\ &\quad + \sum_{r=0}^{7} (\partial^{r+1} f)\,[\sigma_r,\, \zeta] - \sum_{r=0}^{4} (\partial^{r+4} f)\,\xi\sigma_r - \sum_{r=0}^{6} (\partial^{r+2} f)\,\eta\sigma_r\,,\end{aligned}$$

the condition that Z commutes with $\rho_M(f\partial)$ for all $f \in R$ gives

$$\begin{aligned}&[\sigma_r,\, \xi] = 0\,, \quad r = 0, \dots, 7\,,\\ &[\sigma_0,\, \eta] = \eta\,, && [\sigma_0,\, \zeta] = 0\,,\\ &[\sigma_2,\, \eta] = \xi\,, && [\sigma_r,\, \zeta] = \eta\sigma_{r-1}\,, && r = 1, 2\,,\\ &[\sigma_r,\, \eta] = 0\,, \quad r \neq 0, 2\,, && [\sigma_r,\, \zeta] = \eta\sigma_{r-1} + \xi\sigma_{r-3}\,, && r = 3, \dots, 7\,.\end{aligned}$$

We find $\xi = \sigma_5^3$, $\eta = -\sigma_5^2\sigma_3$, $\zeta = \sigma_5^2\sigma_2\sigma_0 + \sigma_5\sigma_3\sigma_2^2 + \sigma_3^4$ in case $(*)$ and $\xi = \sigma_6^3$, $\eta = -\sigma_6^2\sigma_4$, $\zeta = \sigma_6^2\sigma_2\sigma_1 - \sigma_6\sigma_4^2\sigma_1 - \sigma_6\sigma_4\sigma_3\sigma_2 - \sigma_6\sigma_5\sigma_4 - \sigma_4^3\sigma_3$ in case $(**)$. Note that $[Z,\, f_M] = (\partial^3 f)\,\xi + (\partial f)\,\eta$ is not identically zero for all $f \in R$. Hence Z is not a scalar endomorphism, and so M is not absolutely irreducible. In particular, for an absolutely irreducible L_0-module V satisfying either $(*)$ or $(**)$, the $R\mathfrak{D}$-module $\mathcal{R}V$ is no longer absolutely irreducible. It presents no difficulty to ascertain the existence of such modules V.

6. Representations of height one

Although the situation we were dealing with in the previous sections covers many important cases, from the viewpoint of applications to the Lie algebras of vector fields on manifolds it is more natural to consider an arbitrary Lie subalgebra $W \subseteq \operatorname{Der} R$ closed under the R-module structure. In these settings we shall prove results similar to those in Section 4. However, we are far from being able to generalize our previous work completely. In particular, we have to restrict ourselves to the representations which the attribute "height one" may be assigned to.

Put $\Omega^1 = \operatorname{Hom}_R(W, R)$ and define $df \in \Omega^1$ for each $f \in R$ by the rule

$df(D) = Df$, $D \in W$. Our general assumptions for this section are as follows:

(6.1) $2R = R$,

(6.2) as an R-module, W is finitely generated and projective,

(6.3) $\Omega^1 = R \cdot dR$.

Note that (6.2), (6.3) are satisfied for the Lie algebras of all vector fields of respective smoothness class on smooth manifolds, real analytic ones, Stein spaces (cf. [10, Th. 4.1]), the Lie algebras of all polynomial vector fields on smooth affine algebraic varieties. They are also valid when W is generated as an R-module by a finite system of independent derivations. Property (6.1) means, in particular, that we exclude from consideration algebras over fields of characteristic 2.

Our first objective is to introduce a suitable substitute for the category $\mathcal{C}$. Make the R-module $\mathfrak{g} = \Omega^1 \otimes_R W$ into a Lie algebra over R by putting

$$[\theta \otimes D, \theta' \otimes D'] = \langle \theta', D \rangle \, \theta \otimes D' - \langle \theta, D' \rangle \, \theta' \otimes D, \qquad \theta, \theta' \in \Omega^1, \ D, D' \in W.$$

There is a homomorphism σ_W of $\mathfrak{g}$ into the Lie algebra $\mathfrak{gl}_R W$ of all R-linear endomorphisms of W defined by the rule

$$\sigma_W(\theta \otimes D)(D') = -\langle \theta, D' \rangle D, \qquad \theta \in \Omega^1, \quad D, D' \in W.$$

Property (6.2) ensures that σ_W is an isomorphism. For each maximal ideal $\mathfrak{m}$ of R the quotient $\mathfrak{g}/\mathfrak{m}\mathfrak{g}$ is a Lie algebra over the field $R/\mathfrak{m}$ which we identify with the Lie algebra $\mathfrak{gl}(W/\mathfrak{m}W)$ of all linear transformations of the vector space $W/\mathfrak{m}W$ over $R/\mathfrak{m}$. Note that Ω^1, hence also $\mathfrak{g}$ carry W-module structures.

Definition. An irreducible $\mathfrak{g}/\mathfrak{m}\mathfrak{g}$-module is called coexceptional if it is isomorphic to $\bigwedge^t (W/\mathfrak{m}W)^*$ for some $t = 0, \ldots, n$, where $(W/\mathfrak{m}W)^*$ is the dual of the vector space $W/\mathfrak{m}W$ over $R/\mathfrak{m}$.

Definition. Denote by $\mathcal{C}_1$ the category whose objects are additive groups M endowed with a structure of an R-module, a structure ρ_M of a W-module and a structure σ_M of a $\mathfrak{g}$-module so that the following properties are satisfied:

(6.4) $[\rho_M(D), f_M] = (Df)_M$,

(6.5) $[\sigma_M(T), f_M] = 0$,

(6.6) $[\rho_M(D), \sigma_M(T)] = \sigma_M(D \cdot T)$,

(6.7) $\rho_M(fD) = f_M \circ \rho_M(D) + \sigma_M(df \otimes D)$,

(6.8) $\sigma_M(fT) = f_M \circ \sigma_M(T)$

for all $f \in R$, $D \in W$, $T \in \mathfrak{g}$. The morphisms in $\mathcal{C}_1$ are the mappings that respect the three module structures.

In fact (6.6) is a consequence of the other axioms. If $W = R\mathfrak{D}$ where $\mathfrak{D}$ is spanned by an independent system of pairwise commuting derivations then we may identify $\mathcal{C}_1$ with a full subcategory of $\mathcal{C}$ consisting of the objects M

such that $\sigma_M(\widehat{W}_n(k)_1) = 0$. Indeed, the $\widehat{W}_n(k)_0$-module structure on such a M corresponds to a module structure over $\mathfrak{gl}(\mathfrak{D})$, hence also to a module structure over $\mathfrak{gl}_R W \cong R \otimes_k \mathfrak{gl}(\mathfrak{D})$ satisfying (6.8). Properties (6.4)–(6.7) correspond then to (1.1)–(1.4).

If $M, N \in \mathcal{C}_1$ then $M \otimes_R N$ and $\mathrm{Hom}_R(M, N)$ are in a natural way objects of $\mathcal{C}_1$. The category $\mathcal{C}_1$ is closed under arbitrary direct sums and direct products. Our further results concern objects M satisfying the additional assumptions

(6.9) M is a finitely generated projective R-module,

(6.10) for each maximal ideal $\mathfrak{m}$ of R the quotient $M/\mathfrak{m}M$ is an absolutely irreducible module for the Lie $(R/\mathfrak{m})$-algebra $\mathfrak{g}/\mathfrak{m}\mathfrak{g}$,

(6.11) for each maximal ideal $\mathfrak{m}$ of R the $\mathfrak{g}/\mathfrak{m}\mathfrak{g}$-module $M/\mathfrak{m}M$ is not a coexceptional one.

Property (6.10) means that $M \neq \mathfrak{m}M$, in particular. Denote $\mathrm{tr}\, M = \sum \xi(M)$, where the sum is taken over all $\xi \in \mathrm{Hom}_R(M, R)$. Since every linear mapping $M/\mathfrak{m}M \to R/\mathfrak{m}$ admits a lifting to an R-module homomorphism $M \to R$ by projectivity of M, it follows $\mathrm{tr}\, M \not\subseteq \mathfrak{m}$ for each maximal ideal $\mathfrak{m}$. Hence $\mathrm{tr}\, M = R$, i. e. M is a progenerator in the category of R-modules [7, 4.3]. We shall use freely the properties of progenerators.

For $\theta, \theta' \in \Omega^1$, $D, D' \in W$ put

$$\chi_M(\theta, \theta', D, D') = (\sigma_M(\theta \otimes D) - \langle\theta, D\rangle_M)\,\sigma_M(\theta' \otimes D') + (\sigma_M(\theta' \otimes D) - \langle\theta', D\rangle_M)\,\sigma_M(\theta \otimes D').$$

Lemma 6.1. *Let $M \in \mathcal{C}_1$. Then for all $f, g \in R$, $D, D' \in W$ we have*

$$-\rho_M(fgD)\,\rho_M(D') + \rho_M(fD)\,\rho_M(gD') + \rho_M(gD)\,\rho_M(fD') - \rho_M(D)\,\rho_M(fgD') = \chi_M(df, dg, D, D').$$

Proof. By (6.7) and (6.4) we have

$$\rho_M(fD)\,\rho_M(D') - \rho_M(D)\,\rho_M(fD') = (\sigma_M(df \otimes D) - (Df)_M)\,\rho_M(D') - \rho_M(D)\,\sigma_M(df \otimes D').$$

Substituting here first gD for D, then gD' for D', subtracting and applying (6.4)–(6.8), we get the desired formula.

Lemma 6.2. *Let $M = M_1 \oplus \cdots \oplus M_r$ where $M_1, \ldots, M_r$ are objects of $\mathcal{C}_1$ satisfying* (6.9), (6.10). *Suppose that the $(\mathfrak{g}/\mathfrak{m}\mathfrak{g})$-modules $M_i/\mathfrak{m}M_i$, $i = 1, \ldots, r$, are pairwise non-isomorphic for every maximal ideal $\mathfrak{m}$ of R. Then the associative R-subalgebra $B \subseteq \mathrm{End}_R M$ (containing the identity endomorphism) generated by all endomorphisms $\sigma_M(T)$, $T \in \mathfrak{g}$, coincides with $A = \mathrm{End}_R M_1 \times \cdots \times \mathrm{End}_R M_r$.*

Proof. For each maximal ideal $\mathfrak{m}$ of R the image of B in the factor algebra $A/\mathfrak{m}A \cong \prod_{i=1}^r \mathrm{End}_{R/\mathfrak{m}}(M_i/\mathfrak{m}M_i) \subseteq \mathrm{End}_{R/\mathfrak{m}}(M/\mathfrak{m}M)$ is the associative subalgebra generated by all endomorphisms corresponding to the representation of $\mathfrak{g}/\mathfrak{m}\mathfrak{g}$ in $M/\mathfrak{m}M$. By the hypotheses $M/\mathfrak{m}M \cong M_1/\mathfrak{m}M_1 \oplus \cdots \oplus M_r/\mathfrak{m}M_r$ is a semisimple B-module and $\mathrm{End}_B(M/\mathfrak{m}M) \cong K_1 \times \cdots \times K_r$ where $K_i =$

$\mathrm{End}_B(M_i/\mathfrak{m}M_i) \cong R/\mathfrak{m}$ for $i = 1, \ldots, r$. It follows by Jacobson's density theorem that $(B + \mathfrak{m}A)/\mathfrak{m}A \cong \prod_{i=1}^r \mathrm{End}_{K_i}(M_i/\mathfrak{m}M_i) \cong A/\mathfrak{m}A$. Hence $B + \mathfrak{m}A = A$ for each $\mathfrak{m}$. Note that A is finitely generated as an R-module. Hence $B = A$ by Nakayama's lemma.

Lemma 6.3. *Let M be an object of $\mathcal{C}_1$ satisfying* (6.9), (6.10), (6.11). *Then the ideal J of $A = \mathrm{End}_R M$ generated by all endomorphisms $\chi_M(\theta, \theta', D, D')$ with $\theta, \theta' \in \Omega^1$, $D, D' \in W$ coincides with A.*

Proof. For each maximal ideal $\mathfrak{m}$ of R the quotient $J(\mathfrak{m}) = (J + \mathfrak{m}A)/\mathfrak{m}A$ is an ideal of a simple algebra $A/\mathfrak{m}A$. If $J + \mathfrak{m}A = A$ for all $\mathfrak{m}$ then $J = A$ by Nakayama's lemma. Otherwise $J(\mathfrak{m}) = 0$ for at least one $\mathfrak{m}$. Then all endomorphisms $\chi_M(\theta, \theta', D, D')$ induce the zero endomorphism of $M/\mathfrak{m}M$. Denote by $\overline{\sigma}$ the representation of $\mathfrak{g}/\mathfrak{m}\mathfrak{g}$ in $M/\mathfrak{m}M$. We get

$$(\overline{\sigma}(\overline{\theta} \otimes \overline{D}) - \langle \overline{\theta}, \overline{D} \rangle)\, \overline{\sigma}(\overline{\theta'} \otimes \overline{D'}) + (\overline{\sigma}(\overline{\theta'} \otimes \overline{D}) - \langle \overline{\theta'}, \overline{D} \rangle)\, \overline{\sigma}(\overline{\theta} \otimes \overline{D'}) = 0$$

for all $\overline{\theta}, \overline{\theta'} \in (W/\mathfrak{m}W)^*$, $\overline{D}, \overline{D'} \in W/\mathfrak{m}W$. This gives precisely the same relations between the endomorphisms $\overline{\sigma}(T)$, $T \in \mathfrak{g}/\mathfrak{m}\mathfrak{g}$, that we have already encountered in Lemma 4.3, only written in a more invariant form. It follows that $M/\mathfrak{m}M$ is coexceptional, a contradiction.

Lemma 6.4. *Let $M \in \mathcal{C}$, and let M' be a W-submodule of M. Denote by B the subalgebra of the associative R-algebra $\mathrm{End}_R M$ generated by all endomorphisms $\sigma_M(T)$, $T \in \mathfrak{g}$, and put $J = \{\varphi \in B \mid R\varphi(M') \subseteq M'\}$. Then J is an ideal of B containing all endomorphisms $\chi_M(\theta, \theta', D, D')$ with $\theta, \theta' \in \Omega^1$, $D, D' \in W$.*

Proof. It is immediate that J is an R-submodule of B. Being stable under endomorphisms $\rho_M(D)\varphi$ and $\varphi\rho_M(D)$ for all $D \in W$ and $\varphi \in J$, obviously M' is also stable under $\sigma_M(df \otimes D)\varphi = \rho_M(fD)\varphi - \rho_M(D)(f\varphi) + (Df)\varphi$ and similarly under $\varphi\sigma_M(df \otimes D) = \varphi\rho_M(fD) - (f\varphi)\rho_M(D)$, i. e. $\sigma_M(df \otimes D)\varphi \in J$ and $\varphi\sigma_M(df \otimes D) \in J$ for all $f \in R$, $D \in W$, $\varphi \in J$. By (6.3) $\sigma_M(T)\varphi \in J$ and $\varphi\sigma_M(T) \in J$ for all $T \in \mathfrak{g}$, $\varphi \in J$, i. e. J is an ideal of B. Lemma 6.1 shows that M' is stable under endomorphisms $\chi_M(df, dg, D, D')$. Replacing D with hD for $h \in R$, we see that those endomorphisms belong to J. By (6.3) then $\chi_M(\theta, \theta', D, D') \in J$ for all $\theta, \theta' \in \Omega^1$, $D, D' \in W$.

Theorem 6.5. *Let M be an object of $\mathcal{C}_1$ satisfying* (6.9), (6.10), (6.11). *Then the assignment $I \mapsto IM$ establishes a one-to-one correspondence between the W-invariant ideals of R and the W-submodules of M.*

Proof. Put $A = \mathrm{End}_R M$. Let $M' \subseteq M$ be a W-submodule. Assume the notations of Lemma 6.4. Then $B = A$ by Lemma 6.2, and $J = A$ by Lemma 6.3. In other words, M' is an A-submodule. The assignment $I \mapsto IM$ establishes a one-to-one correspondence between the ideals of R and the A-submodules of M [7, Th. 4.7]. Hence $M' = IM$ for some I. Let $D \in W$. For $f \in I$, $m \in M$ we have $\rho_M(D)(fm) \equiv (Df)m \pmod{M'}$ according to (6.4). Hence $\rho_M(D)M' + M' = (DI + I)M$. Now $DI + I$ is an ideal of R, and we see that $\rho_M(D)M' + M'$ is the corresponding A-submodule of M. Hence $DI \subseteq I$ if and only if $\rho_M(D)M' \subseteq M'$.

We conclude that the property of W-invariance for M' is equivalent to that for I.

Remark. Obviously, W with its natural R-module structure, the adjoint representation on itself and the representation σ_W already introduced is an object of $\mathfrak{C}_1$. Under assumption that the ideal of R generated by all elements Df with $D \in W$, $f \in R$ coincides with R but without (6.2), (6.3) the description of ideals of W was obtained earlier [9], [10], [12]. Projectivity or a close reproduction property (see [10]) is needed, however, to ensure the bijectivity of the correspondence between the W-invariant ideals of R and the ideals of W. In fact, for specific modules our approach allows to obtain results under assumptions weaker than those imposed presently.

Theorem 6.6. *Let M, N be two objects of $\mathfrak{C}_1$ satisfying* (6.9), (6.10). *Assume that either M or N satisfies* (6.11) *as well. Then any W-module homomorphism $\xi : M \to N$ is a morphism in $\mathfrak{C}_1$. If the $(\mathfrak{g}/\mathfrak{m}\mathfrak{g})$-modules $M/\mathfrak{m}M$ and $N/\mathfrak{m}N$ are non-isomorphic for all maximal ideals $\mathfrak{m}$ of R then necessarily $\xi = 0$. If $M = N$ then* $\operatorname{End}_W M \cong R^W$.

Proof. Put $A_M = \operatorname{End}_R M$, $A_N = \operatorname{End}_R N$, $A = A_M \times A_N$. Denote by B the associative R-subalgebra of A (containing unity) generated by all pairs $(\sigma_M(T), \sigma_N(T))$, $T \in \mathfrak{g}$. Let $p_M : A \to A_M$ and $p_N : A \to A_N$ be the canonical projections. By Lemma 6.2 $p_M(B) = A_M$ and $p_N(B) = A_N$. Put $J = \{(\varphi, \psi) \in B \mid \xi \circ f_M \circ \varphi = f_N \circ \psi \circ \xi \text{ for all } f \in R\}$. In other words, J consists of all pairs $(\varphi, \psi) \in B$ such that the graph $\Gamma(\xi)$ of ξ is stable under endomorphisms $(f\varphi, f\psi)$ for all $f \in R$. By Lemma 6.4 J is an ideal of B containing the pairs $(\chi_M(\theta, \theta', D, D'), \chi_N(\theta, \theta', D, D'))$ for all $\theta, \theta' \in \Omega^1$, $D, D' \in W$. As $p_M(J)$ and $p_N(J)$ are ideals of A_M and A_N respectively, Lemma 6.3 shows that $p_M(J) = A_M$ or $p_N(J) = A_N$ depending on which of the modules M, N satisfies (6.11). Suppose, for instance, the first equality holds. Then $(1_M, \psi) \in J$ for some $\psi \in A_N$. For all $f \in R$ we have $\xi f_M = f_N \psi \xi$. In particular, $\xi = \psi \xi$, whence $\xi f_M = f_N \xi$. It follows $(1_M, 1_N) \in J$, i. e. $J = B$. The same is valid when $p_N(J) = A_N$. We conclude $(\sigma_M(T), \sigma_N(T)) \in J$, whence $\xi \sigma_M(T) = \sigma_N(T) \xi$ for all $T \in \mathfrak{g}$. Thus ξ is a morphism in $\mathfrak{C}_1$. Under assumption that $M/\mathfrak{m}M$ and $N/\mathfrak{m}N$ are non-isomorphic $(\mathfrak{g}/\mathfrak{m}\mathfrak{g})$-modules for all maximal ideals $\mathfrak{m}$ of R we deduce from Lemma 6.2 that $B = A$, hence $J = A$. It follows $(0_M, 1_N) \in J$, and so $\xi = 1_N \xi = \xi 0_M = 0$. Suppose $M = N$. Then $B = \{(\varphi, \varphi) \mid \varphi \in A_M\}$. Hence ξ is an endomorphism of M as an A_M-module. By [7, Cor. 4.2] $\operatorname{End}_{A_M} M \cong R$. Hence $\xi = f_M$ for some $f \in R$. According to (6.4) we have also $(Df)_M = 0$, and so $Df = 0$ for all $D \in W$, i. e. $f \in R^W$.

Definition. An object $P \in \mathfrak{C}_1$ is called invertible if it is invertible as an R-module, i. e. if there exists an R-module Q such that $P \otimes_R Q \cong R$.

Theorem 6.7. *Let M, N be two objects of $\mathfrak{C}_1$ satisfying* (6.9), (6.10). *Suppose that the Jacobson radical* $\operatorname{Jac} R$ *of R contains no nonzero W-invariant ideals of R. If the $(\mathfrak{g}/\mathfrak{m}\mathfrak{g})$-modules $M/\mathfrak{m}M$ and $N/\mathfrak{m}N$ are isomorphic for all maximal* *ideals* $\mathfrak{m}$ *of* R, *then* $N \cong P \otimes_R M$ *in* $\mathcal{C}_1$ *for a suitable invertible object* $P \in \mathcal{C}_1$ *with* $\sigma_P = 0$. *If in addition* M, N *satisfy* (6.11) *then* $\mathrm{Hom}_R(M, N) \cong P^W$.

Proof. We use the notations A_M, A_N, B introduced in the proof of Theorem 6.6. We regard M and N as B-modules via the projections $p_M : B \to A_M$ and $p_N : B \to A_N$. The kernel K of p_M is an ideal of B. Hence $p_N(K)$ is an ideal of A_N, and it follows that $p_N(K) = IA_N$ for a suitable ideal I of R. For each maximal ideal $\mathfrak{m}$ of R the quotients $M/\mathfrak{m}M$ and $N/\mathfrak{m}N$ are also B-modules. By the hypotheses there is a $(\mathfrak{g}/\mathfrak{m}\mathfrak{g})$-module isomorphism $\eta : M/\mathfrak{m}M \to N/\mathfrak{m}N$. Since B is generated by $(\sigma_M(T), \sigma_N(T))$, $T \in \mathfrak{g}$, it is clear that η is a B-module isomorphism as well. Now K annihilates M, hence also $M/\mathfrak{m}M$ and $N/\mathfrak{m}N$. It follows that IA_N annihilates $N/\mathfrak{m}N$, i. e. $IA_N \subseteq \mathfrak{m}A_N$. Since the bijective correspondence between the ideals of R and those of A_N preserves the inclusions [7, Th. 4.7], we deduce $I \subseteq \mathfrak{m}$. As the inclusion is valid for all maximal ideals $\mathfrak{m}$ of R, it follows $I \subseteq \mathrm{Jac}\, R$. There are canonical W-module structures on A_M, A_N. In view of (6.6) $([\rho_M(D), \sigma_M(T)], [\rho_N(D), \sigma_N(T)]) = (\sigma_M(D \cdot T), \sigma_N(D \cdot T)) \in B$ for all $D \in W$, $T \in \mathfrak{g}$, i. e. B is a W-invariant subalgebra of $A_M \times A_N$. As the projections are W-invariant, K and also IA_N are stable under W. If $D \in W$ then $I + DI$ is an ideal of R. Since $(Df)\varphi = [\rho_M(D), f\varphi] - f[\rho_M(D), \varphi] \in IA_N$ for all $f \in I$, $\varphi \in A_N$ we have $(I + DI)A_N = IA_N$, whence $I + DI = I$. It follows that I is W-invariant. By the hypotheses $I = 0$. Hence $K = 0$, i. e. p_M is an isomorphism of B onto A_M. By symmetry p_N is an isomorphism as well.

By Morita theory the functors $S \mapsto S \otimes_R M$ from the category of R-modules to the category of left B-modules and $U \mapsto \mathrm{Hom}_B(M, U)$ in the opposite direction are inverse to each other up to natural isomorphisms. Thus, if we put $P = \mathrm{Hom}_B(M, N)$, then the mapping $\nu : P \otimes_R M \to N$, $\xi \otimes m \mapsto \xi(m)$ for $\xi \in P$, $m \in M$, is an isomorphism of B-modules. An immediate verification shows that $\rho_N(D)\xi - \xi\rho_M(D) \in P$ for all $D \in W$, $\xi \in P$. Hence there is a W-module structure on P compatible with the R-module structure. Since $\sigma_N(df \otimes D)\xi = \xi\sigma_M(df \otimes D)$, we have $\rho_N(fD)\xi - \xi\rho_M(fD) = f(\rho_N(D)\xi - \xi\rho_M(D))$ for all $f \in R$, $D \in W$, $\xi \in P$. Thus we may regard P as an object of $\mathcal{C}_1$ with $\sigma_P = 0$. By symmetry there is an isomorphism of B-modules $M \cong Q \otimes_R N$ for an appropriate R-module Q. Then $M \cong Q \otimes_R P \otimes_R M$ as B-modules, whence $Q \otimes_R P \cong R$ as R-modules. In other words, P is invertible.

If M, N satisfy (6.11), then any W-module homomorphism $\xi : M \to N$ is a morphism in $\mathcal{C}_1$. In particular, $\xi \in P$. Then obviously $\xi \in P^W$. The proof is completed.

We conclude with an explicit construction of modules. Assume that R is an algebra over a field of characteristic zero. For each positive integer r the r-fold tensor product $W^{\otimes r}$ over R is an object of $\mathcal{C}_1$ satisfying (6.9). The symmetric group $\mathfrak{S}_r$ operates via automorphisms of this object. Let $k\mathfrak{S}_r$ denote the group algebra of $\mathfrak{S}_r$ and S_λ its simple module corresponding to a partition $\lambda = (\lambda_1, \ldots, \lambda_s)$ of r. Put

$$W^\lambda = \mathrm{Hom}_{k\mathfrak{S}_r}(S_\lambda, W^{\otimes r}).$$

The three module structures on $W^{\otimes r}$ induce those on W^λ. It follows that $W^\lambda \in$

$\mathcal{C}_1$. Since every representation of $\mathfrak{S}_r$ is completely reducible, we have $W^{\otimes r} \cong \oplus(W^\lambda \otimes_k S_\lambda)$, the sum being taken over all partitions. In particular, W^λ is isomorphic with a direct summand of $W^{\otimes r}$, hence satisfies (6.9). For each maximal ideal $\mathfrak{m}$ of R

$$W^\lambda/\mathfrak{m}W^\lambda \cong \mathrm{Hom}_{k\mathfrak{S}_r}(S_\lambda, (W/\mathfrak{m}W)^{\otimes r}).$$

According to Weyl's description of irreducible modules for the general Lie algebra of linear transformations, $W^\lambda/\mathfrak{m}W^\lambda$ vanishes for $s > \dim_{R/\mathfrak{m}} W/\mathfrak{m}W$ and is an absolutely irreducible $(\mathfrak{g}/\mathfrak{m}\mathfrak{g})$-module otherwise. Note also that $W^\lambda/\mathfrak{m}W^\lambda$ is not coexceptional. In the context of Lie algebras of vector fields W^λ is the module of contravariant tensor fields of type λ. If we start with Ω^1 instead of W we can define similarly objects $\Omega^\lambda \in \mathcal{C}_1$ corresponding to covariant vector fields of type λ. Now $\Omega^\lambda/\mathfrak{m}\Omega^\lambda$ is coexceptional for partitions having the form $(1,\ldots,1)$.

7. Exceptional height one modules

We keep the assumptions of Section 6. Furthermore, we assume that

W is projective of constant rank n as an R-module.

This means that $\dim W/\mathfrak{m}W = n$, hence also $\dim \Omega^1/\mathfrak{m}\Omega^1 = n$ for all maximal ideals $\mathfrak{m}$ of R. Denote $\Omega^t = \bigwedge_R^t \Omega^1$, $A^t = \mathrm{End}_R\,\Omega^t$. Then $\Omega^t = 0$ for $t < 0$ and $t > n$. Otherwise Ω^t is a progenerator in the category of R-modules. Hence the functor $P \mapsto \Omega^t P = \Omega^t \otimes_R P$ from the category of R-modules to that of left A^t-modules is a category equivalence. The elements of $\Omega^t P$ are called P-valued differential forms of degree t. For an R-module homomorphism $\xi : P' \to P$ denote by $\xi_* : \Omega^t P' \to \Omega^t P$ the A^t-module homomorphism afforded by the functoriality. If P' is a submodule of P then we identify $\Omega^t P'$ with a submodule of $\Omega^t P$.

For $D \in W$ let i_D denote the endomorphism of inner product on the exterior algebra $\Omega = \bigwedge_R \Omega^1$, which we extend also to $\Omega P = \Omega \otimes_R P$. Given $\theta \in \Omega^1$ and $D \in W$, define $\sigma^t(\theta \otimes D) \in A^t$ by the rule

$$\sigma^t(\theta \otimes D)(\omega) = \theta \wedge i_D(\omega), \qquad \omega \in \Omega^t.$$

Then $\sigma^t : \mathfrak{g} \to A^t$ is a Lie algebra homomorphism. Hence $\mathfrak{g}$ operates on $\Omega^t P$ for every P.

We assume further that P is an object of $\mathcal{C}_1$ with $\sigma_P = 0$. Then ΩP carries additional structures of a W-module and of a complex with respect to the exterior differential $d = d_P$ (see, for example, [22, sect. 3]). Denote by ρ the representation of W in ΩP. The classical formula

(7.1) $\rho(D) = d\,i_D + i_D\,d$

shows that $\big(\rho(fD) - f\,\rho(D)\big)\omega = df \wedge i_D(\omega) = \sigma^t(df \otimes D)\,\omega$ for $f \in R$, $D \in W$, $\omega \in \Omega^t P$. Hence $\Omega^t P \in \mathcal{C}_1$. If P is invertible then $\Omega^t P$ satisfies (6.9), (6.10) but the $(\mathfrak{g}/\mathfrak{m}\mathfrak{g})$-modules $\Omega^t P/\mathfrak{m}\Omega^t P$ are coexceptional. As is well known, $\Omega^t P$

contains W-invariant subspaces of closed differential forms Z^tP and the exact ones B^tP which are not R-submodules in general. Hence the results of Section 6 are no longer valid. Denote by H^tP the cohomology group Z^tP/B^tP.

Lemma 7.1. *Let P' be a W-invariant R-submodule of P. Then*

(i) *if $t > 0$ then $\Omega^t P'$ is spanned by the differential forms $\theta \wedge \omega$ with $\theta \in \Omega^1$, $\omega \in \Omega^{t-1}P'$;*

(ii) *if $t < n$ then $\Omega^t P'$ is spanned by the differential forms $i_D(\omega)$ with $D \in W$, $\omega \in \Omega^{t+1}P'$;*

(iii) *if $t < n$, $\omega \in \Omega^t P$ and $\theta \wedge \omega \in \Omega^{t+1}P'$ for all $\theta \in \Omega^1$ then $\omega \in \Omega^t P'$;*

(iv) *if $t > 0$, $\omega \in \Omega^t P$ and $i_D(\omega) \in \Omega^{t-1}P'$ for all $D \in W$ then $\omega \in \Omega^t P'$;*

(v) *if $t > 0$ then $\Omega^t P' = R \cdot B^t P'$;*

(vi) *if $t < n$, $\omega \in \Omega^t P$ and $d(R\omega) \subseteq \Omega^{t+1}P'$ then $\omega \in \Omega^t P'$;*

(vii) *if $t < n$ then $\Omega^t P'$ is spanned by the differential forms $i_D\, d(\omega)$ with $D \in W$, $\omega \in \Omega^t P'$;*

(viii) *if $t > 0$, $\omega \in \Omega^t P$ and $d\, i_D(\omega) \in \Omega^t P'$ for all $D \in W$ then $\omega \in \Omega^t P'$;*

Proof. (i)–(iv) are consequences of the projectivity of W.

We have $df \wedge \omega = d(f\omega) - f\, d\omega$ for all $f \in R$, $\omega \in \Omega P$. Hence (v) follows from (i). Similarly, (vi) follows from (iii).

Since $i_D(f\omega) = i_{fD}(\omega) = f\, i_D(\omega)$, (vii) follows from (v) and (ii); similarly, (viii) follows from (vi) and (iv).

Theorem 7.2. *Let U be a W-submodule of $\Omega^t P$, $0 \le t \le n$. Then there are two uniquely determined W-invariant R-submodules P', $P'' \subseteq P$ such that $P'' \subseteq P'$ and*

$$d(\Omega^{t-1}P') + \Omega^t P'' \subseteq U \subseteq \Omega^t P' \cap d^{-1}(\Omega^{t+1}P''),$$

where we set $P' = P$ for $t = 0$ and $P'' = 0$ for $t = n$. Conversely, given two W-invariant R-submodules P', $P'' \subseteq P$, any additive subgroup $U \subseteq \Omega^t P$ satisfying the above inclusions is stable under W.

Proof. We first describe (R, W)-submodules of $\Omega^t P$. Any such submodule is stable under the endomorphisms $\sigma^t(df \otimes D)$ with $f \in R$, $D \in W$ in view of (6.7). It is also stable under the associative R-subalgebra generated by those endomorphisms which coincides with A^t by Lemma 6.2. It remains to notice that the assignment $Q \mapsto \Omega^t Q$ establishes a one-to-one correspondence between the R-submodules of P and the A^t-submodules of $\Omega^t P$ and that the W-invariant R-submodules correspond to the W-invariant A^t-submodules.

Denote $J = \{(\varphi, \psi) \in A^t \times A^{t+1} \mid (d\, i_{D'}\, \varphi + i_{D'}\, \psi\, d)(U) \subseteq U$ for all $D' \in W\}$. By (7.1) J contains the pair of identity endomorphisms (id, id). If $(\varphi, \psi) \in J$, $f \in R$ then U is stable under $d\, i_{fD'}\, \varphi + i_{fD'}\, \psi\, d = d\, i_{D'}\, f\varphi + i_{D'}\, f\psi\, d$. Hence J is an R-submodule. Let $(\varphi, \psi) \in J$, $f \in R$, D, $D' \in W$. Then U is stable under endomorphisms

$$\rho(D)(d\, i_{D'}\, \varphi + i_{D'}\, \psi\, d) - (d\, i_{[D,D']}\, \varphi + i_{[D,D']}\, \psi\, d) = d\, i_{D'}\, \rho(D)\, \varphi + i_{D'}\, \rho(D)\, \psi\, d$$

and

$$(d i_{D'} \varphi + i_{D'} \psi d)\rho(D) = d i_{D'} \varphi \, \rho(D) + i_{D'} \psi \, \rho(D) \, d,$$

hence also under

$$\begin{aligned} d i_{D'} \rho(fD) \varphi + i_{D'} \rho(fD) \psi d - (d i_{fD'} \rho(D) \varphi + i_{fD'} \rho(D) \psi d) \\ = d i_{D'} \sigma^t (df \otimes D) \varphi + i_{D'} \sigma^{t+1} (df \otimes D) \psi d \end{aligned}$$

and

$$\begin{aligned} d i_{D'} \varphi \, \rho(fD) + i_{D'} \psi \, \rho(fD) \, d - (d i_{D'} \varphi f \, \rho(D) + i_{D'} \psi f \, \rho(D) \, d) \\ = d i_{D'} \varphi \sigma^t (df \otimes D) + i_{D'} \psi \sigma^{t+1} (df \otimes D) \, d. \end{aligned}$$

Therefore, both $\big(\varphi \sigma^t (df \otimes D),\ \psi \sigma^{t+1}(df \otimes D)\big)$ and $\big(\sigma^t(df \otimes D)\varphi,\ \sigma^{t+1}(df \otimes D)\psi\big)$ are contained in J. By Lemma 6.2 the pairs $\big(\sigma^t(df \otimes D),\ \sigma^{t+1}(df \otimes D)\big)$ generate the R-algebra $A^t \times A^{t+1}$. Hence J is its ideal, and so $J = A^t \times A^{t+1}$. In particular, J contains (id, 0) and (0, id). This means that U is stable under endomorphisms $d i_D$ and $i_D d$ for all $D \in W$.

Obviously, $U' = RU$ and $U'' = (U : R) = \{\omega \in U \mid R\omega \subseteq U\}$ are (R, W)-submodules of $\Omega^t P$. By the remark at the beginning of the proof $U' = \Omega^t P'$ and $U'' = \Omega^t P''$ for suitable W-invariant R-submodules P', $P'' \subseteq P$. In order to ensure the uniqueness we alter, however, the definition of P' and P'' respectively for $t = 0$ and $t = n$ as indicated in statement of the theorem. The inclusions $U'' \subseteq U \subseteq U'$ are immediate. In particular, it follows $P'' \subseteq P'$. As we have proved, $d i_D(RU) \subseteq U$ for all $D \in W$. Hence $d(\Omega^{t-1} P') \subseteq U$ by Lemma 7.1 (ii). We have proved also $R i_D d(U) \subseteq U$, i. e. $i_D d(U) \subseteq U''$ for all $D \in W$. By Lemma 7.1 (iv) $dU \subseteq \Omega^{t+1} P''$. The first assertion is proved.

Conversely, let P', P'' be given, and let an additive subgroup U satisfy the stated inclusions. If $\omega \in U$ then $i_D(\omega) \in \Omega^{t-1} P'$ and $d\omega \in \Omega^{t+1} P''$, hence $D\omega \in d(\Omega^{t-1} P') + \Omega^t P'' \subseteq U$ for all $D \in W$ according to (7.1). By Lemma 7.1 (v) $RU = \Omega^t P'$ provided $t > 0$. By Lemma 7.1 (vi) $(U : R) = \Omega^t P''$ provided $t < n$. This shows that P', P'' are uniquely determined with respect to U. The proof is completed.

Corollary 7.3. *If P is a simple object of $\mathfrak{C}_1$ (in particular, if R is W-simple and P is invertible) then any W-submodule $U \subseteq \Omega^t P$ satisfy $U = 0$, $U = \Omega^t P$ or $B^t P \subseteq U \subseteq Z^t P$.*

Proof. The only possibilities for (P'', P') are $(0, 0)$, (P, P) and $(0, P)$.

Theorem 7.4. *Let P, Q be two objects of $\mathfrak{C}_1$ with $\sigma_P = 0$, $\sigma_Q = 0$. Let $\zeta : \Omega^t P \to \Omega^{t'} Q$ be a W-module homomorphism, $0 \le t, t' \le n$.*

(i) *If $t = t'$ then $\zeta = \xi_*$, where $\xi : P \to Q$ is a morphism in $\mathfrak{C}_1$.*

(ii) *If $t + 1 = t'$ then $\zeta = \xi_* d_P = d_Q \xi_*$, where $\xi : P \to Q$ is a morphism in $\mathfrak{C}_1$.*

(iii) *If $t \ne t'$, $t + 1 \ne t'$ and either $t \ne n$ or $t' \ne 0$ then $\zeta = 0$.*

(iv) *If $t = n > 1$, $t' = 0$ then ζ is the composite of the canonical projection*

$\Omega^n P \to H^n P$, a k-linear mapping $H^n P \to H^0 Q$ and the canonical embedding $H^0 Q \to \Omega^0 Q$.

Proof. In the associative R-algebra $A^t \times A^{t+1} \times A^{t'} \times A^{t'+1}$ consider the subalgebra B generated by all quadruples $\big(\sigma^t(T), \sigma^{t+1}(T), \sigma^{t'}(T), \sigma^{t'+1}(T)\big)$ with $T \in \mathfrak{g}$. By Lemma 6.2 B is isomorphic to the direct product of the components A^{t_i} corresponding to different t_i among t, $t+1$, t', $t'+1$. The graph $\Gamma(\zeta)$ of ζ is obviously a W-submodule of $\Omega^t P \oplus \Omega^{t'} Q$. Denote by J the subset of all quadruples $(\varphi, \psi, \varphi', \psi') \in B$ such that $\Gamma(\zeta)$ is stable under the endomorphisms $(d i_D \varphi + i_D \psi d,\ d i_D \varphi' + i_D \psi' d)$ of $\Omega^t P \oplus \Omega^{t'} Q$ for all $D \in W$, i. e. such that

$$(d i_D \varphi' + i_D \psi' d)\,\zeta = \zeta\,(d i_D \varphi + i_D \psi d) \qquad \text{for all} \quad D \in W.$$

It is verified quite similarly as in the proof of Theorem 7.2 that J is an ideal of B containing the quadruple of identity endomorphisms (id, id, id, id). Hence $J = B$. We shall use the invariance of ζ with respect to various components of B.

(i) Here $\Gamma(\zeta)$ is identified with a W-submodule of $\Omega^t(P \oplus Q)$. By Theorem 7.2

$$d(\Omega^{t-1}\Gamma') + \Omega^t \Gamma'' \subseteq \Gamma(\zeta) \subseteq \Omega^t \Gamma' \cap d^{-1}(\Omega^{t+1}\Gamma''),$$

where Γ', Γ'' are W-invariant R-submodules of $\Gamma(\zeta)$. Let $p : P \oplus Q \to P$ and $q : P \oplus Q \to Q$ be the canonical projections. Applying $p_* : \Omega^t(P \oplus Q) \to \Omega^t P$, we get

$$d(\Omega^{t-1}P') + \Omega^t P'' \subseteq \Omega^t P = p_*\Gamma(\zeta) \subseteq \Omega^t P' \cap d^{-1}(\Omega^{t+1}P''),$$

where $P' = p(\Gamma')$, $P'' = p(\Gamma'')$. By the uniqueness part of Theorem 7.2 we get $P'' = P$ provided $t < n$. Then p_* maps $\Omega^t\Gamma''$ onto $\Omega^t P$. Since p_* induces an isomorphism of $\Gamma(\zeta)$ onto $\Omega^t P$, it follows $\Omega^t \Gamma'' = \Gamma(\zeta)$. It follows also that p maps Γ'' isomorphically onto P. Hence Γ'' is the graph of an (R, W)-module homomorphism $\xi : P \to Q$ and $\zeta = \xi_*$. Note that ξ is a morphism in $\mathcal{C}_1$. Assume now that $t = n$. Then Theorem 7.2 ensures $P' = P$. Denote by K the kernel of the restriction $p|_{\Gamma'}$. Then $d(\Omega^{t-1}K) \subseteq \Gamma(\zeta) \cap \ker p_* = 0$. By Lemma 7.1 (vi) $\Omega^{t-1}K = 0$, i. e. $K = 0$. Thus Γ' is the graph of an (R, W)-module homomorphism $\xi : P \to Q$, i. e. a morphism in $\mathcal{C}_1$. Then $\Omega^t\Gamma'$ is the graph of ξ_*. Since $\Gamma(\zeta) \subseteq \Omega^t\Gamma'$, we deduce $\Gamma(\zeta) = \Omega^t\Gamma'$, i. e. $\zeta = \xi_*$.

(ii) Here $B \cong A^t \times A^{t'} \times A^{t'+1}$. The invariance of ζ with respect to $A^{t'}$ yields $\zeta i_D d = d i_D \zeta$ for all $D \in W$. Put $U = \{(\omega, \omega') \in \Omega^t P \oplus \Omega^t Q \mid \zeta(\omega) = d(\omega')\}$ and $U' = \{\big(d(\omega), \zeta(\omega)\big) \mid \omega \in \Omega^t P\}$. We identify U, U' with W-submodules of $\Omega^t(P \oplus Q)$, $\Omega^{t'}(P \oplus Q)$ respectively. By Theorem 7.2 $d(\Omega^t\Gamma') \subseteq U' \subseteq RU' = \Omega^{t'}\Gamma'$ and $\Omega^t\Gamma'' = (U : R) \subseteq U \subseteq d^{-1}(\Omega^{t'}\Gamma'')$ for suitable W-invariant R-submodules $\Gamma', \Gamma'' \subseteq P \oplus Q$. Let $K = \ker p|_{\Gamma''}$. Then $\Omega^t K \subseteq U \cap \ker p_*$, whence $d(\Omega^t K) = 0$ by the definition of U, and $K = 0$ by Lemma 7.1 (vi). Let $P' = p(\Gamma')$. Then $p_*(U') \subseteq \Omega^{t'} P'$. On the other hand, $p_*(U') = d(\Omega^t P)$ by the definition of U'. Hence $P' = P$ by Lemma 7.1 (vi). We have $i_D(U') \subseteq U$ for all $D \in W$ because this is equivalent to the equality $\zeta i_D d = d i_D \zeta$. Since $f \circ i_D = i_{fD} = i_D \circ f$ for $f \in R$, $D \in W$, we have $i_D(RU') \subseteq (U : R)$ for all $D \in W$. Hence $\Gamma' \subseteq \Gamma''$

by Lemma 7.1 (iv). It follows that $\Gamma' = \Gamma''$ and p maps Γ' isomorphically onto P. Hence Γ' is the graph of a morphism $\xi : P \to Q$ in $\mathcal{C}_1$. It is immediate that $\zeta = d\xi_* = \xi_* d$.

(iii), (iv) The invariance of ζ with respect to A^{t+1} yields $\zeta i_D d = 0$ for all $D \in W$. It follows by Lemma 7.1 (vii) $\zeta = 0$ unless $t = n$. The invariance of ζ with respect to $A^{t'}$ yields $d i_D \zeta = 0$ for all $D \in W$. It follows by Lemma 7.1 (viii) $\zeta = 0$ unless $t' = 0$. Assume $t = n > 1$, $t' = 0$. Then the invariance of ζ with respect to A^t yields $\zeta d i_D = 0$ for all $D \in W$, whence ζ vanishes on $B^n P$ by Lemma 7.1 (ii). The invariance of ζ with respect to $A^{t'+1}$ yields $i_D d\zeta = 0$ for all $D \in W$, whence the image of ζ is contained in $Z^0 Q$ by Lemma 7.1 (iv).

References

[1] Blattner R. J., Induced and produced representations of Lie algebras, Trans. Amer. Math. Soc. 144 (1969) 457–474.

[2] Chang H. J., Über Wittsche Lie-Ringe, Abh. Math. Univ. Hamburg 14 (1941) 151–184.

[3] Dzhumadil'daev A. S., Irreducible representations of strongly solvable Lie algebras over a field of positive characteristic (Russian), Mat. Sbornik 123 (1984) 212–229.

[4] Dzhumadil'daev A. S., Central extensions of the Zassenhaus algebra and their irreducible representations (Russian), Mat. Sbornik 126 (1985) 473–489.

[5] Dzhumadil'daev A. S., Cohomologies of truncated coinduced representations of Lie algebras of positive characteristic (Russian), Mat. Sbornik 180 (1989) 456–468.

[6] Ermolaev Yu. B., On irreducible modules with a filtration ($p > 0$) (Russian), Izv. Vyssh. Uchebn. Zaved. Mat. no. 7 (1981) 80–84.

[7] Faith C., Algebra: Rings, Modules and Categories I, Springer-Verlag, Berlin - Heidelberg - New York, 1973.

[8] Farnsteiner R., Strade H., Shapiro's lemma and its consequences in the cohomology theory of modular Lie algebras, Math. Z. 206 (1991) 153–168.

[9] Grabowski J., Isomorphisms and ideals of the Lie algebras of vector fields, Invent. Math. 50 (1978) 13–33.

[10] Grabowski J., Ideals of the Lie algebras of vector fields revisited, Rend. Circ. Mat. Palermo 32 (1993) 89–95.

[11] Jordan D. A., Simple Lie rings of derivations of commutative rings, J. London Math. Soc. (2) 18 (1978) 443–448.

[12] Jordan D. A., On the ideals of a Lie algebra of derivations, J. London Math. Soc. 33 (1986) 33–39.

[13] Koreshkov N. A., On irreducible representations of the Lie p-algebra W_n (Russian), I, VINITI no. 376-82; II, VINITI no. 377-82 (1982).

[14] Koreshkov N. A., On irreducible representations of maximal subalgebras of Cartan type Lie algebras (Russian), Izv. Vyssh. Uchebn. Zaved. Mat. no. 11 (1986) 26–32.

[15] Kostrikin A. I., Shafarevich I. R., Graded Lie algebras of finite characteristic (Russian), Izv. Akad. Nauk USSR Ser. Mat. 33 (1969) 251–322.

[16] Krylyuk Ya. S., Algebras of Cartan type: representations and prolongations (Russian), Moscow State Univ., Cand. Diss., 1978.

[17] Kuznetsov M. I., Truncated induced modules over transitive Lie algebras of characteristic p (Russian), Izv. Akad. Nauk USSR Ser. Mat. 53 (1989) 557–589.

[18] Mil'ner A. A., The irreducible representations of the Zassenhaus algebra (Russian), Uspekhi Mat. Nauk 30 no. 6 (1975) 178.

[19] Rudakov A. N., Irreducible representations of the infinite dimensional Lie algebras of Cartan type (Russian), Izv. Akad. Nauk USSR Ser. Mat. 38 (1974) 835–866.

[20] Skryabin S. M., On representations of Lie algebras of derivations of commutative rings (Russian), Vestnik Moscow Univ. Ser. I. Mat. Mekh. no. 4 (1987) 32–35.

[21] Skryabin S. M., Modular Lie algebras of Cartan type over algebraically non-closed fields. I, Comm. Algebra 19 (1991) 1629–1741.

[22] Skryabin S. M., An algebraic approach to the Lie algebras of Cartan type, Comm. Algebra 21 (1993) 1229–1336.

[23] Strade H., Representations of the Witt algebra, J. Algebra 49 (1977) 595–605.

Theory of invariants and Galois modules

*V. E. Voskresenskiĭ**

1. At first I shall consider a problem which was very interesting for N. G. Chebotarev, and the method of its solution opened a new page in the theory of algebraic varieties.

Let k be a field of characteristic 0, $F = k(x_1, \dots, x_n)$ be a field of rational functions of n variables. Let G be a group acting faithfully on F/k by transformations of variables. Is the field of invariants F^G always rational over k? This is a famous question of E. Noether. The answer is negative in the general case. It was proved for an unclosed field k and an abelian group G [11, 12, 13], and for a closed field in [10].

Let me us recall how this can be done in the case of the field of constants $k = \mathbb{Q}$ and a cyclic group G acting on the set $\{x_1, \dots, x_n\}$ by cyclic transformation. The group G can be represented as a subgroup of the general linear group $\mathrm{GL}_{\mathbb{Q}}(n)$ and, since G is commutative, G is a subgroup of a maximal $\mathbb{Q}$-torus S of $\mathrm{GL}_{\mathbb{Q}}(n)$, $\dim S = n$. Thus,

$$F^G = \mathbb{Q}(S/G) = \mathbb{Q}(T),$$

where $T = S/G$ is a $\mathbb{Q}$-torus. The problem is reduced to rationality of the torus T over $\mathbb{Q}$, and in the theory of algebraic tori there are enough tools for clarifying problems of rationality. By the way, the torus T is the spectrum of the ring of invariants. More exactly, let $\mathfrak{G}$ be the Galois group of the extension $\bar{\mathbb{Q}}/\mathbb{Q}$, and let $\hat{T}$ be the $\mathfrak{G}$-module of rational characters of T. Then $T = \operatorname{Spec} A$, where $A = (\bar{\mathbb{Q}}[\hat{T}])^{\mathfrak{G}}$. In our case, when G is a cyclic group of order n, the ring A can be represented in the following form:

$$A = L[\hat{T}]^{\Pi}, \Pi = \mathrm{Gal}(L/\mathbb{Q}), L = \mathbb{Q}(\zeta_n), |\Pi| = \phi(n).$$

The problem of rationality of the field F^G is reduced to the problem of rationality for another field of invariants $L(\hat{T})^{\Pi}$, where Π acts nontrivially on L and on $\hat{T}$. The problem of the birational structure of the torus T over $\mathbb{Q}$ can be clarified by considering a smooth projective model X of T. The open embedding $T \subset X$ defines the flasque resolution

$$0 \longrightarrow \hat{T} \longrightarrow \hat{M} \longrightarrow \operatorname{Pic} \bar{X} \longrightarrow 0 \tag{1}$$

of $\hat{T}$, and the $\mathfrak{G}$-module $\operatorname{Pic} \bar{X}$ determines the variety T up to stable equivalence,

*Supported by a grant from Russian Ministry of Education.

$\bar{X} = X \otimes \bar{\mathbb{Q}}$. Thus we come to a problem from another domain, that is, we have to deal with Galois modules of finite ranks and their invariants.

2. I shall give another very significant example. Let $M(n) = M(n, \mathbb{C})$ be the ring of square matrices of order n with complex coefficients, $V_{m,n} = M(n) \oplus \cdots \oplus M(n)$ is the direct sum of m copies of the space $M(n)$. The group $\mathrm{GL}(n, \mathbb{C})$ acts on $V_{m,n}$ by componentwise conjugation:

$$(B_1, \ldots, B_m) \longrightarrow (AB_1A^{-1}, \ldots, AB_mA^{-1}),$$

where $B_i \in M(n)$, $A \in \mathrm{GL}(n, \mathbb{C})$, $m \geq 2$, $n \geq 2$. We denote the categorical quotient with respect to this action by $V_{m,n}/\mathrm{GL}_n$. The field of rational functions on the quotient $V_{m,n}/\mathrm{GL}_n$ (the field of invariants) will be denoted by $K_{m,n}$. The quotient $V_{m,n}/\mathrm{GL}_n$ is birationally equivalent to the direct product $V_{2,n}/\mathrm{GL}_n \times M(n)^{m-2}$. Thus, the field $K_{m,n}$ is a pure transcendental extension of the field $K_{2,n}$. The problem of rationality of the field $K_{2,n}$ has been studied for a long time. It amounts to classifying pairs of matrices up to simultaneous conjugation. Rationality of the field $K_{2,2}$ was shown by Silvester but after that the work on this problem stopped. At last Formanek showed that the field $K_{2,n}$ can be realized as the field of invariants $\mathbb{C}(U_n \oplus V_n)^{S_n}$, where S_n is the symmetric group acting by permutations on the basis $e_1, \ldots, e_n$, U_n is the lattice $\langle e_1, \ldots, e_n \rangle$, and V_n is the module obtained from the exact sequence

$$0 \longrightarrow V_n \longrightarrow U_n \otimes U_n \xrightarrow{\phi} U_n \xrightarrow{\varepsilon} \mathbb{Z} \longrightarrow 0,$$

where $\varepsilon(e_i) = 1$, $\phi(e_i \otimes e_j) = e_i - e_j$. We see that the field $K_{2,n}$ is a field of rational functions on the F_n-torus T_n, where F_n is the field of invariants $\mathbb{C}(U_n)^{S_n}$, and $\hat{T}_n$ is V_n. Formanek's investigations [6, 7] showed that the fields $K_{2,3}$ and $K_{2,4}$ are rational over $\mathbb{C}$. Recently it was shown in the work [1] that the field $K_{2,n}$ is stable rational for $n = 5, 7$, and this result shows that $K_{2,n}$ are stable rational for all $n \leq 420$.

3. A very similar construction appears in problems of birational classification of semisimple groups. Let G be a connected semisimple algebraic group defined over a field of characteristic 0, T a maximal k-torus of the group G, $N = N_G(T)$ the normalizer of the torus T in G, and $W = N/T$ the Weyl group. The variety of maximal tori $X = G/N$ is rational over k; see [2]. Let $H \subset X \times_k G$ be a closed subset given by

$$H = \{ (x, g) \mid gx = x \}.$$

Then the first projection defines the morphism $\pi : H \to X$ and $\pi^{-1}(x)$ is a maximal torus of the group $G \times_k k(x)$. Since the variety H is birationally equivalent to G over K we have $k(G) = F(H_x)$, where $F = k(x)$, H_x is a fibre over a generic point of the scheme X, and H_x is a generic torus of the group G. The construction of the generic torus is known at least for simply connected and adjoint semisimple groups [14]. Thus it is possible to compute the birational invariant $H^1(F, \operatorname{Pic} \bar{Y}(H_x))$ for such groups. It is equal to 0, see [8]. So we can describe the arithmetic of maximal tori without affect. Using this fact Kunyavskii and Skorobogatov gave an elegant proof of weak approximation in adjoint and simply connected groups [9].

4. Let us return to the sequence (1). The study of its properties led us to an algebraic construction of the flasque resolution of a finitely generated torsion-free Π-module $\hat{T}$, where Π is any finite group. It is an exact sequence

$$0 \longrightarrow \hat{T} \longrightarrow \hat{S} \longrightarrow \hat{N} \longrightarrow 0 \tag{2}$$

where $\hat{S}$ is a permutation module and $H^{-1}(\pi, \hat{N}) = 0$ for all subgroups π of the group Π. The sequence (2) defines the module $\hat{N}$ up to similarity. The modules $\hat{N}$ and $\hat{M}$ are called similar if there exist permutation modules $\hat{S}_1$ and $\hat{S}_2$ such that $\hat{N} \oplus \hat{S}_1 \cong \hat{M} \oplus \hat{S}_2$. The class of similarity $[\hat{N}]$ of the module $\hat{N}$ depends only on $\hat{T}$. We denote it by $p_\Pi(\hat{T})$ or $p(\hat{T})$. The importance of this invariant is explained by the theory of algebraic varieties. Let $C(\Pi)$ be the category of finitely generated torsion-free Π-modules, L/K a Galois extension with the Galois group Π, and $C(L/K)$ the category of algebraic tori defined over k and split over L. The class $p_\Pi(\hat{T})$ is a birational invariant of the torus T. It determines the variety T up to stable equivalence. The class $p_\pi(\hat{T})$ is also a birational invariant of the tori T. We have a set of invariants $\{p_\pi(\hat{T}) \mid \pi \subset \Pi\}$ for a given Π-module $\hat{T}$. We would like to have an algorithm which allows to distinguish quickly modules with trivial or non-trivial invariants $p_\pi(\hat{T})$. In the next section we shall consider this problem for the Π-modules of the special form but these modules appear as first obstruction in the study of more complicated cases.

We shall give some standard notation. Let $\mathbb{Z}[\Pi/\Gamma]$ be the module generated by classes $g\Gamma$ with the natural action of the group Π, $\mathbb{Z}[\Pi/\Gamma] = \mathbb{Z}[\Pi] \otimes \mathbb{Z}$, $\epsilon : \mathbb{Z}[\Pi/\Gamma] \to \mathbb{Z}$ is an epimorphism, $\epsilon(\Sigma a_g(g\Gamma)) = \Sigma a_g$; $N : \mathbb{Z} \to \mathbb{Z}[\Pi/\Gamma]$ is the norm map, $N(1) = \Sigma g\Gamma$. Let $I_{\Pi/\Gamma} = \ker(\epsilon)$, $J_{\Pi/\Gamma} = \mathbb{Z}[\Pi/\Gamma]/\mathbb{Z}$, $I^0_{\Pi/\Gamma} = J_{\Pi/\Gamma}$, $M^0 = \mathrm{Hom}(M, \mathbb{Z})$. Always we have $p_\Pi(I_{\Pi/\Gamma}) = 0$, but this isn't always true for the module $J_{\Pi/\Gamma}$.

5. Lattices in the space $\mathbb{Q}(\zeta_n)$. Let $\mathbb{Z}[\zeta_n]$ be the ring of integral elements of the cyclotomic field $\mathbb{Q}(\zeta_n)$, where ζ_n is a primitive root of unity of n-th degree. It is a root of the over $\mathbb{Q}$ irreducible cyclotomic polynomial $\Phi_n(x)$. The rank of the lattice $\mathbb{Z}[\zeta_n]$ is equal to $\phi(n)$, where ϕ is Euler function. Let Γ_n be the cyclic group of order n with a generator σ. Define the structure of the Γ_n-module on $\mathbb{Z}[\zeta_n]$ by putting $\sigma(\alpha) = \zeta_n\alpha$. Then all ideals $\mathfrak{P}$ of the ring $\mathbb{Z}[\zeta_n]$ will also be Γ_n-modules. They describe tori with cyclic fields of decomposition up to stable equivalence. Let Δ_n be the Galois group of the extension $\mathbb{Q}(\zeta_n)/\mathbb{Q}$. The group Δ_n is isomorphic to the multiplicative class group modulo n, and its order is equal to $\phi(n)$. The action of the group Δ_n on $\mathbb{Z}[\zeta_n]$ is determined by its action on the element ζ_n. Operators of Γ_n and Δ_n generate the group Π_n, and Π_n is semidirect product of a cyclic normal subgroup Γ_n and an abelian group Δ_n, i.e., $\Pi_n = \Gamma_n\Delta_n$. The module $\mathbb{Z}[\zeta_n]$ is a lattice in the $\mathbb{Q}$-space $\mathbb{Q}(\zeta_n)$,which we can consider as Π_n-module. It will be natural to describe all Π_n-modules M of finite rank in $\mathbb{Q}(\zeta_n)$ such that $\mathbb{Q}M = \mathbb{Q}(\zeta_n)$. Let M be such a module. Since M is a Γ_n-module, M is

a fractional ideal of the field $\mathbb{Q}(\zeta_n)$, with the additional condition $\tau M = M$ for all $\tau \in \Delta_n$. It is clear that we may consider M as an integral ideal. Let $\mathfrak{P}$ be a prime ideal dividing M. Then $\tau(\mathfrak{P}) \mid M$, $\tau \in \Delta_n$. Let $\mathfrak{O}(\mathfrak{P})$ be a product of different prime ideals which lie in the orbit of the ideal $\mathfrak{P}$. If the ideal $\mathfrak{P}$ is unramified then $\mathfrak{O}(\mathfrak{P}) = (p)$, where p is a prime, and the ideals M and pM are isomorphic as Π_n-modules. Let d be the discriminant of $\mathbb{Q}(\zeta_n)$, $p_i \mid d$, $(p_i) = \mathfrak{O}\mathfrak{P}_i)^{e_i}$, $e_i > 1$.

Proposition 1. *All Π_n-modules M of finite rank with the condition $M \otimes \mathbb{Q} \cong \mathbb{Q}(\zeta_n)$ are, up to Π_n-isomorphism, among the ideals of the ring $\mathbb{Z}[\zeta_n]$ of the form*

$$\prod_{\mathfrak{P}|p_i|d} \mathfrak{O}(\mathfrak{P}_i)^{a_i}, \tag{3}$$

where $0 \le a_i < e_i$, $e_i > 1$, and for every p_i we take only one factor $\mathfrak{P}_i$. The modules of the form (3) *are pairwise non-isomorphic .*

The uniqueness of the representation (3) is a consequence of the absolute irreducibility of the Π_n-module $\mathbb{Q}(\zeta_n)$ and, hence, of the modules of the form (3). We are interested in rationality of tori corresponding to modules of the form (3).

It is suitable to consider the crossed product Σ_n of the group Δ_n and the field $\mathbb{Q}(\zeta_n)$ for further investigations. The algebra Σ_n is generated over the field $\mathbb{Q}(\zeta_n)$ by the basis

$$u_\tau,\ \tau \in \Delta_n,\ u_{\tau\omega} = u_\tau u_\omega,\ \alpha u_\tau = u_\tau(\tau(\alpha)),\ \alpha \in \mathbb{Q}(\zeta_n).$$

Although the algebra Σ_n is a trivial crossed product, this construction is very useful. It is easy to prove that Σ_n is a factor-algebra $\mathbb{Q}[\Pi_n]/(\Phi_n(\sigma))$, where $(\Phi_n(\sigma))$ is a principal ideal in the group ring $\mathbb{Q}[\Pi_n]$. The factor-ring $\mathbb{Z}[\Pi_n]/(\Phi_n(\sigma)) = \Lambda_n$ is a maximal order in the $\mathbb{Q}$-algebra Σ_n. As Π_n-module the order Λ_n can be represented in the form

$$\Lambda_n \cong \mathbb{Z}[\Pi_n] \otimes_\Gamma \mathbb{Z}[\zeta_n].$$

Since the birational invariant $p_{\Gamma_n}(\mathbb{Z}[\zeta_n]) = [0]$, we have $p_{\Pi_n}(\Lambda_n) = [0]$. On the lattice $\mathbb{Z}[\zeta_n]$ we can consider a bilinear integral form $(\alpha, \beta) = \operatorname{tr}(\alpha\bar{\beta})$. It is symmetric, nondegenerate and invariant with respect to the group Π_n. The form (α, β) gives a possibility to compute the module $M^0 = \operatorname{Hom}(M, \mathbb{Z})$, which is dual to the module M. At first $\mathbb{Z}[\zeta_n]^0 = D^{-1}$ where D is the different of the field $\mathbb{Q}(\zeta_n)$. Let M be an ideal of the form (3), M^{-1} the inverse ideal for M, and

$$M^* = \{\, \alpha \in \mathbb{Q}(\zeta_n) \mid (\alpha, \beta) \in \mathbb{Z} \quad \forall \beta \in M \,\}.$$

Then

$$M^0 \cong M^* = D^{-1}M^{-1}. \tag{4}$$

Let us consider the special case where $n = p$ is a prime. In this case the group Δ_p is also cyclic of order $p-1$, the group Π_p is metacyclic of order $p(p-1)$. Only one prime ramifies in the field $\mathbb{Q}(\zeta_p)$, namely $(p) = \mathfrak{P}^{p-1}$, where $\mathfrak{P}$ is a principal ideal, $\mathfrak{P} = (\zeta_p - 1)$, $N\mathfrak{P} = p$. Modules of the form (3) in $\mathbb{Q}(\zeta_p)$ are exhausted by the ideals

$$\mathbb{Z}[\zeta_p] = (1),\ \mathfrak{P},\ \mathfrak{P}^2, \dots,\ \mathfrak{P}^{p-2}.$$

The ideal $\mathfrak{P}^{p-2}$ is the different of the field $\mathbb{Q}(\zeta_p)$. The isomorphism $(\mathfrak{P}^i)^0 \cong \mathfrak{P}^{p-1}$ follows from (4). In particular, $\mathbb{Z}[\zeta_n]^0 \cong \mathfrak{P}$. The polynomial $\Phi_p(\sigma)$ has the form $1+\sigma+\cdots+\sigma^{p-1}$ and the Π_p-module $\mathbb{Z}[\zeta_p]$ can be represented as

$$\mathbb{Z}[\zeta_p] \cong \mathbb{Z}[\Pi/\Delta]/\mathbb{Z} = J_{\Pi/\Delta},\ \Pi = \Pi_p.$$

Thus $\mathfrak{P} \cong J^0_{\Pi/\Delta} \cong I_{\Pi/\Delta}$ and, hence, $p_\Pi(\mathfrak{P}) = [0]$. In the article [5] the following isomorphism of Π-modules was established:

$$\Lambda_p \cong \mathbb{Z}[\zeta_p] \oplus \mathfrak{P} \oplus \cdots \oplus \mathfrak{P}^{p-2}.$$

We know that in this correlation $p_\Pi(\Lambda_p) = p_\Pi(\mathfrak{P}) = [0]$. It is useful to study the filtration

$$(1) \supset \mathfrak{P} \supset \mathfrak{P}^2 \supset \cdots \supset \mathfrak{P}^{p-2}.$$

The factor-modules $\mathfrak{P}^i/\mathfrak{P}^{i+1}$ are isomorphic as groups to the cyclic group $\mathbb{Z}/p\mathbb{Z}$, but the structures of the Π_p-modules on these factors are different. Let

$$\pi = (\zeta_p - 1),\ \tau(\pi) = \zeta_p^a - 1,\ \langle\tau\rangle = \Delta_p.$$

Proposition 2. *The group Γ_p acts trivially on the factors $\mathfrak{P}^i/\mathfrak{P}^{i+1}$. The Π_p-module $\mathfrak{P}^i/\mathfrak{P}^{i+1}$ can be identified with the Δ_p-module $\mathbb{Z}/p\mathbb{Z}$ such that τ acts on it according to the rule $\tau(1) \equiv a^i \pmod{p}$.*

We shall use these facts for quite concrete computations. Let $p = 5$. The group Π_5 has order 20. Π_5 is a semi-simple product of the group $\Gamma_5 = \langle\sigma\rangle$ of order 5 and the cyclic group $\Delta_5 = \langle\tau\rangle$ of order 4, $\tau\sigma\tau^{-1} = \sigma^2$, $\tau(\zeta_5) = \zeta_5^2$,

$$\Lambda_5 \cong \mathbb{Z}[\zeta_5] \oplus \mathfrak{P} \oplus \mathfrak{P}^2 \oplus \mathfrak{P}^3. \tag{5}$$

$$\mathfrak{P} = (\pi),\ \mathfrak{P}^4 = (5),\ N\mathfrak{P} = 5.$$

Let us describe the tori corresponding to the modules $\mathfrak{P}^i$ explicitly. Let $k = \mathbb{Q}$, and let $a \in \mathbb{Q}$ such that the polynomial $x^5 - a$ is irreducible over $\mathbb{Q}$. Let $F = \mathbb{Q}(\sqrt[3]{a})$, $L = F(\zeta_5) = \mathbb{Q}(\sqrt[5]{a}, \zeta_5)$. To the module $\mathbb{Z}[\zeta_5]$ corresponds the norm hypersurface $R^1_{F/\mathbb{Q}}(\mathbb{G}_m)$. To the ideal $\mathfrak{P}$ corresponds the torus $R_{F/\mathbb{Q}}(\mathbb{G}_m)/\mathbb{G}_m$. It is an open domain in $\mathbb{P}^4_{\mathbb{Q}}$, hence, $p_\Pi(\mathfrak{P}) = [0]$. The torus corresponding to the ideal $\mathfrak{P}^2$ is birationally equivalent over $\mathbb{Q}$ to the quotient of $\mathbb{P}^4_{\mathbb{Q}}$ by the cyclic group $\mathbb{Z}/5\mathbb{Z}$ such that its action on $\mathbb{P}^4$ is determined by linear action over $\mathbb{Q}$ in the five-dimensional space $R_{F/\mathbb{Q}}(\mathbb{G}_m)$ by the operator of multiplication on ζ_5 in matrix form. Since the quotient $R_{F/\mathbb{Q}}(\mathbb{G}_m)/(\mathbb{Z}/5\mathbb{Z})$ is rational over $\mathbb{Q}$, it follows that $\mathbb{P}^4/(\mathbb{Z}/5\mathbb{Z})$ is stable rational (in fact, it is rational) and, consequently, $p_\Pi(\mathfrak{P}^2) = [0]$. The equality (5) shows that $p_\Pi(\mathbb{Z}[\zeta_5]) + p_\Pi(\mathfrak{P}^3) = [0]$. In the work [3] it is shown that the element $P_\Pi(\mathbb{Z}[\zeta_5])$ has infinite order in the group of stable equivalence. Hence, $p_\Pi(\mathfrak{P}^3) = -p_\Pi(\mathbb{Z}[\zeta_5])$ is also distinct from zero.

Possibly, there is an observable rule in for the birational classification of the modules $\mathfrak{P}^i$ for $p > 5$.

6. Let us consider in more detail the birational invariant $p(\hat{T}) = p(T) = [\hat{N}]$, defined by the flasque resolution (2). Let $S(\Pi)$ be the class of all permutations

Π-modules of finite rank.

$$H^q(\Pi) = \{ \hat{M} \in C(\Pi) \mid H^q(\pi, \hat{M}) = 0 \forall \pi \subset \Pi \},$$

where H^q is Tate cohomology. Then, let

$$D(\Pi) = \{ \hat{M} \in C(\Pi) \mid \hat{M} \oplus \hat{M}' \in S(\Pi) \}.$$

We have the inclusions

$$S(\Pi) \subset D(\Pi) \subset H(\Pi) \subset H^{\pm 1}(\Pi) \subset C(\Pi),$$

where $H(\Pi) = H^{-1}(\Pi) \cap H^1(\Pi)$. Classes of similarity form commutative semigroups with respect to the direct sum of modules. We denote them

$$D(\Pi)/S(\Pi) \subset H(\Pi)/S(\Pi) \subset H^{\pm 1}(\Pi)/S(\Pi) \subset C(\Pi/S(\Pi).$$

The classes $D(\Pi)/S(\Pi)$ form a maximal subgroup in the semigroup $C(\Pi)/S(\Pi)$. The Picard class $p(T)$ lies in $H^{-1}(\Pi)/S(\Pi)$ and the map $T \to p(T)$ defines the isomorphism

$$p : Z(L/k) \to H^{-1}(\Pi)/S(\Pi),$$

where $Z(L/k)$ is the semigroup of tori of the category $C(L/k)$ up to stable equivalence [13].

Let us return to our case. The group $\Pi = \Pi_P$ is metacyclic, that means that all its subgroups are cyclic. Then, as it is known, $P_\Pi(\hat{M}) \in D(\Pi)/S(\Pi)$ for any Π-module $\hat{M} \in C(\Pi_P)$. The group $D(\Pi)/S(\Pi)$ has a finite number of generators but it isn't always a finite group.

Proposition 3. *Let p be a prime, $p \geq 5$. Then the class $p_\Pi(\mathbb{Z}[\zeta_p])$ has infinite order in the group $D(\Pi)/S(\Pi)$.*

Proof. The order of the group Π_P is equal to $p(p-1)$ and $p-1 \geq 4$. Therefore, either $p-1 = 2^\alpha$, $\alpha \geq 2$, or there exists an odd prime q, $q \mid p-1$. Let $A = \Delta_p$ in the first case and let A be a subgroup of order q in the group Δ_p in the second case. The semidirect product $\pi = \Gamma_P A$ is a subgroup in Π_P. The action of the group A on Γ_P by conjugation is faithful, $\omega\sigma\omega^{-1} = \sigma^k$, k has order $|A|$ in the multiplicative group of numbers modulo p. In [4] it is shown that the class $p_\pi(\mathfrak{P}^0)$ has infinite order in the group $D(\pi)/S(\pi)$. Since $\mathfrak{P}^0 = \mathrm{Hom}(\mathfrak{P}, \mathbb{Z}) \cong \mathbb{Z}[\zeta_p]$, we conclude that $p_\pi(\mathbb{Z}[\zeta_p])$ and, hence, $p_\Pi(\mathbb{Z}[\zeta_p])$ has infinite order.

References

[1] Bessenrodt C., Le Bruyn L., Stably rationality of certain GL(n) quotients, Invent. Math. 104 (1991), 179–199.

[2] Chevalley C. On algebraic group varieties, J.Math.Soc.Japan. 6 (1954), 303–324.

[3] Colliot-Thèlène J.-L., Sansuc J.-J., La R-equivalence sur les tores, Ann. Scient. Ec. Norm. Super. 10 (1977), 175–230.

[4] Endo S., Miyata T., On a classification of the function fields of algebraic tori, Nagoya Math. J. 56 (1975), 85–104.

[5] Endo S., Miyata T., Sekuguchi K., Picard Groups and Automorphism Groups of integral Group Rings of Metacyclic Groups, J. Algebra 77 (1982), 286–310.

[6] Formanek E., The center of 3 × 3 generic matrices, Linear and Multilinear Algebra 7 (1979), 203–212.

[7] Formanek E., The center of 4 × 4 generic matrices, J. Algebra. 7 (1980), 304–319.

[8] Klyachko A. A., Tori without affect in semisimple groups, Arithmetic and Geometry of varieties, Kuibyshev Univ. (1989), 67–78 (Russian).

[9] Kunyavskii B. E., Skorobogatov A. N., Weak Approximation in Algebraic Groups and Homogeneous Spaces, Contemp. Math. 131 (1991), 447–451.

[10] Saltman D. J., Noether's problem over an algebraically closed fields, Invent. Math. 77 (1984), 71–84.

[11] Swan R. G., Invariant rational functions and a problem of Steenrod, Invent. Math. 7 (1969), 148–158.

[12] Voskresenskii V. E., On the Birational Equivalence of Linear Algebraic Groups, Soviet. Math. Dokl. 10 (1969), 1212–1215.

[13] Voskresenskii V. E., Algebraic tori, Nauka, Moscow, 1977 (Russian).

[14] Voskresenskii V. E., Maximal tori without affect in semisimple algebraic groups, Math. Notes 44 (1988), 651–655.

List of invited speakers

Dmitri V. Alekseevsky
Gen. Antonova 2, kv. 99
117279 Moscow
Russia

S. Barry Cooper
School of Mathematics
University of Leeds
Leeds LS2 9JT
England

E. P. Dolzhenko
Department of Mathematics
Moscow State University
Vorob'evu goru
119899, Moscow
Russia

A. S. Dzhumadil'daev
Department of Algebra
Institute of Mathematics
Puschkin str. 125
480021 Almaata
Kazachstan

V. Ja. Gutljanski
Institute of Applied Mathematics and Mechanics
Universitetskaja 77
340048 Donezk
Ukraine

Helmut Koch
Max-Planck-Arbeitsgruppe
"Algebraische Geometrie und Zahlentheorie"
Jägerstraße 10 – 11
10117 Berlin
Germany

A. I. Kostrikin
Department of Mathematics
Moscow State University
Vorob'evu goru
119899, Moscow
Russia

Vic. S. Kulikov
Moscow State University
of Transport Communications
(MIIT)
ul. Obraztsova 15
101475 Moscow, A-55
Russia

I. R. Shafarevich
Steklov Institute of Mathematics
Vavilova 42
117966 Moscow
Russia

T. Shioda
Department of Mathematics
Rikkyo University
Nishi-Ikebukuro, Tokyo 171
Japan

V. E. Voskresenskii
Department of Mathematics
Samara State University
Avrora str. 157, 12
443067 Samara
Russia

List of contributors

Dmitri V. Alekseevsky
Gen. Antonova 2, kv. 99
117279 Moscow
Russia

Walter L. Baily, Jr .
University of Chicago
Department of Mathematics
5734 University Ave.
Chicago, IL 60637-1514
USA

V. A. Chugunov
Department of Applied Mathematics
Kazan State University
Lenin str. 18
420008 Kazan
Russia

S. Barry Cooper
School of Mathematics
University of Leeds
Leeds LS2 9JT
England

L. D. Eskin
Department of Applied Mathematics
Kazan State University
Lenin str. 18
420008 Kazan
Russia

David A. Jordan
School of Mathematics and Statistics
University of Sheffield
The Hicks Building
Sheffield S3 7RH
United Kingdom

Helmut Koch
Max-Planck-Arbeitsgruppe
"Algebraische Geometrie und Zahlentheorie"
Jägerstraße 10 – 11
10117 Berlin
Germany

A. I. Kostrikin
Department of Mathematics
Moscow State University
Vorob'evu goru
119899, Moscow
Russia

Vic. S. Kulikov
Moscow State University
of Transport Communications
(MIIT)
ul. Obraztsova 15
101475 Moscow, A-55
Russia

M. I. Kuznetsov
Department of Mechanics and Mathematics
Nizhny Novgorod State University
603600 Nizhny Novgorod
Russia

I. R. Shafarevich
Steklov Institute of Mathematics
Vavilova 42
117966 Moscow
Russia

T. Shioda
Department of Mathematics
Rikkyo University
Nishi-Ikebukuro, Tokyo 171
Japan

Serge M. Skryabin
Institute of Mathematics and Mechanics
Kazan State University
Universitetskaja 17
420008 Kazan
Russia

Andrea F. Spiro
Dipartimento di Matematica "V. Volterra"
Via Brecce Bianche
University of Ancona
60131 Ancona
Italy

S. L. Tonkonog
Department of Applied Mathematics
Kazan State University
Lenin str. 18
420008 Kazan
Russia

V. E. Voskresenskii
Department of Mathematics
Samara State University
Avrora str. 157, 12
443067 Samara
Russia

Bei Fragen zur Produktsicherheit wenden Sie sich bitte an:
If you have any questions regarding product safety,
please contact:

Walter de Gruyter GmbH
Genthiner Straße 13
10785 Berlin
productsafety@degruyterbrill.com